TRANSACTIONS

*of the*

American Philosophical Society

*Held at Philadelphia for Promoting Useful Knowledge*

VOLUME 82, Part 3

# The Preadamite Theory and the Marriage of Science and Religion

**David N. Livingstone**
School of Geosciences
The Queen's University of Belfast

Library of Congress Catalog
Card Number 91-76986
International Standard Book Number 0-87169-823-4
US ISSN 0065-9746

To the Memory
of
Alan Flavelle

# CONTENTS

# LIST OF ILLUSTRATIONS

# Acknowledgments

I am grateful to Charles C. Gillispie for helpful comments on an earlier draft of this monograph; to Arie Leegwater for translations from the Dutch and for insightful observations on the topic; to Ronald L. Numbers for suggesting some useful lines of inquiry; to Mark A. Noll for his suggestions on the relationship between monogenism and moral philosophy; to Ellen Alderink for assistance with the figures; and to several of my students—in particular, Paul Overvoorde, Bräm Weidenaar, and Deborah Stout—for their various contributions in discussions of the subject. A travel grant from the Royal Society allowed me to consult materials held in the library of the American Philosophical Society, Philadelphia.

# Introduction

An altogether just war has been waged, of late, by historians on the so-called warfare interpretation of the relations between science and religion. As they have ransacked the documents of the scientific past, historians of science have dismantled the conflict model with forensic precision.[1] But while the inadequacy of the military metaphor has been patiently exposed, and a clarion call issued for the writing of "non-violent" histories, surprisingly little attention has been paid to those schemes, developed by scientists and theologians alike, that were specifically designed to merge scriptural revelation and scientific information. Social and cognitive histories of, say, the "gap theory" which posited a lengthy period of time between the first two verses of the book of Genesis, or of the "day age theory" which interpreted the creatorial days as geological epochs, or indeed of the "days of revelation theory" according to which the Genesis creation story refers to successive days of divine disclosure, might well direct our attention to those who conceived of the relationship between science and religion as dialectical rather than independent or antagonistic.[2]

Among the harmonizing tactics that have been deployed to

---

[1] This is illustrated in the following recent works: David C. Lindberg and Ronald L. Numbers, eds, *God and Nature. Historical Essays on the Encounter between Christianity and Science* (Berkeley: University of California Press, 1986); idem, "Beyond War and Peace: A Reappraisal of the Encounter between Christianity and Science," *Church History*, 55 (1986): 338-354; James R. Moore, *The Post-Darwinian Controversies: A Study of the Protestant Struggle to Come to Terms with Darwin in Great Britain and America, 1870-1900* (Cambridge: Cambridge University Press, 1979); David N. Livingstone, *Darwin's Forgotten Defenders: The Encounter between Evangelical Theology and Evolutionary Thought* (Edinburgh: Scottish Academic Press, 1987). A useful review of the literature on the warfare thesis is available in Ronald L. Numbers, "Science and Religion," *Osiris*, 2nd series, 1 (1985): 59-80.

[2] The gap theory was advanced by Thomas Chalmers in "Remarks on Cuvier's Theory of the Earth; in Extracts from a Review of that Theory which was Contributed to The Christian Instructor in 1814," *The Select Works of Thomas Chalmers* (New York: Robert Carter and Brothers, 1850), vol. 1, pp. 180-193. The day age theory was supported by Hugh Miller, *The Testimony of the Rocks; or, Geology in Its Bearing on the Two Theologies, Natural and Revealed* (Edinburgh: William P. Nimmo, 1856). The idea that the days might refer to six daily visions is outlined in P.J. Wiseman, *Clues to Creation in Genesis* (London: Marshall, Morgan and Scott, 1958); it was earlier put forward by Jesuits, S.J. Hummelauer in his *Commentary on Genesis* and by P. Méchineau in *Historicité de Genese*, and criticized by M.J. Lagrange, in "Hexameron (Genese 1 a 2, 4)," *Revue Biblique*, 5 (1896): 391. Various harmonizing strategies are reviewed in Ronald L. Numbers, *Creation by Natural Law. Laplace's Nebular Hypothesis in American Thought* (Seattle and London: University of Washington Press, 1977).

keep alive the marriage of science and religion, is the beguilingly simple preadamite theory—the idea that human beings existed before the biblical Adam.[3] Indeed it is perhaps the very simplicity of the idea that has enabled it to retain a coterie of committed defenders over the years and to survive in the recesses of the Western mind for centuries, periodically resurfacing to serve different social and cognitive functions. The recently published survey of the influence of Isaac de la Peyrère's preadamite theory by Richard Popkin has opened up certain aspects of the early history of the scheme; but the continuing vitality of preadamism in specific strands of twentieth century theology that were beyond Popkin's scrutiny, not to mention neglected aspects of its nineteenth-century manifestation, means that we remain without a basic guidebook to its overall historical dimensions. It is therefore my aim here to identify some of the major conceptual landmarks in the changing preadamite territory up to the present day.

Of course there are myriad dangers involved in the attempt to take in several centuries of intellectual history at a bite, and there are bound to be historical infelicities that experts in specific eras will all too easily identify. I apologize for them in advance. Yet I have judged such risks worth taking in order that an outline map of the terrain may become available. For only when a preliminary survey has been carried out can the extraordinary vitality of the preadamite theory, and its versatility in serving scientific, religious, and social interests, begin to be appreciated. It is my suspicion that the history of other harmonizing models that have been advanced might be scrutinized profitably by historians of science and religion, not least in a day when the boundary line between science and religion has been shown to be far more fluid—if indeed it exists at all—than previous generations have thought.

It is therefore my hope that this overview of the preadamite theory may reveal how interpenetrative scientific and religious discourses have been, and that by investigating one variety of harmonizing scheme, some of the more general characteristics of the species may be identified.

---

[3] A preliminary survey of the theory's history is available in David N. Livingstone, "Preadamites: The History of an Idea from Heresy to Orthodoxy," *Scottish Journal of Theology*, 40 (1987): 41-66. See also idem, "Preadamism: the History of a Harmonizing Strategy," *Fides et Historia*, 22 (1990): 25-34.

# I. Preadamism as Skepticism

While the modern history of preadamism can be traced to the impact of the so-called voyages of exploration on European culture and the revival of classical learning during the Renaissance, fleeting glimpses of the preadamite are detectable among earlier writers. To be sure, these versions of the theory appear in altogether esoteric philosophical and metaphysical contexts that exert little influence on the subsequent history of the idea. But their lingering presence was sufficient to keep the memory of alternative world chronologies subliminally flickering in the western mind. Accordingly we will turn first to the early history of preadamism, bearing in mind that these early contexts differed markedly from the social and cognitive environments within which it later flourished.

## Preadamism and Pagan Chronology

In the early Christian era there was, understandably, a clash between the cosmologies of the Judaeo-Christian world and those of other cultures. Around 170 A.D., for example, Theophilus of Antioch felt the need to contest the claim of Apollonius the Egyptian that the world was 153,075 years old,[1] while Julian the Apostate (c. 331-363), the Roman emperor who attempted to restore pagan learning, is said to have advocated preadamism.[2] Again, Gregory of Nyssa (330-c.395) is reputed to have argued the case for Adam's physical body being derived from animal forebears, an understandable view perhaps in the light of the outlook of his mentor Origen (c.185-c.254) whose concept of the infusion of pre-existent souls into human bodies was all of a piece with his conviction that Adam's physical form was only given to him as a consequence of the fall. As for Gregorian anthropology itself, it emphasized humanity's continuity *and* discontinuity with the rest of the natural order, inasmuch as the distinctively human *rational* soul was superadded to the *vegetative* and *sensitive* (or

---

[1] See Paul H. Kocher, *Christopher Marlowe, A Study of His Thought, Learning, and Character* (New York: Russell & Russell, 1962), 44.

[2] A.J. Mass, "Préadamites," in *The Catholic Encyclopedia. An International Work of Reference on the Constitution, Doctrine, Discipline and History of the Catholic Church* (New York: The Encyclopedia Press, 1913), vol. 12, s.v.

1

*animated*) souls—elements belonging to the plant and animal or-
ders respectively—to produce the rational human animal that
was a blend of every soul-type. But this did not preclude human
emergence from animal antecedents. Because he believed every-
thing existed in spermatic potential from the initial divine impulse of creation, Gregory could and did advance a developmentalist account of the origin of life-forms and urged that the human
body had been created through the inherent activity of the elements of the earth.[3] And so it is not surprising that in later centuries Christian evolutionists—Catholics in particular—would
look back to Gregory of Nyssa to find legitimacy within the Christian tradition for their evolutionary proposals.

Whether these doctrines are strictly preadamite is perhaps an
open question. But they certainly contributed to a tradition,
which included Augustine, that approached the question of human origins in developmentalist terms. By the same token they
also contributed to the skeptical ethos that brought Augustine to
the point of refuting the "abbominable lyings" of those Egyptians
who claimed that the world was a hundred thousand years old.[4]
Beyond these, traces of preadamism are also present in the Mi-
drash and Cabalistic literature where it was connected to the notion of the plurality of worlds. The idea here was that this world
is merely the latest in a sequence of divine creations which had
been destroyed due to God's disapproval; some even speculated
that this possibility is hinted at in the Genesis narrative by the fact
that the text begins with a *Beth* instead of an *Aleph*. And yet other
strains are discernible too. *Nabatean Agriculture*, for example, an
apologia for Babylonian society against Islam that appeared in
A.D. 904 suggested that Adam had come from India to be the
progenitor, not of the human race, but of an agricultural civilization. This scheme was later reported in the *Guide for the Perplexed*

---

[3] See E.C. Messenger, *Evolution and Theology* (London: Burns, Oates and Washburne,
1931); Henry de Dorlodot, *Darwinism and Catholic Thought*, translated by E. C. Messenger
(London: Burns, Oates & Co., 1922); Christian Pesch, *De Deo Creante et Elevante, de Deo Fine
Ultimo Tractatus Dogmatici* (Friburgi Brisgoviae: Herder, 1909); Adolf Harnack, "Origen,"
*Encyclopaedia Britannica* 14th edition (London and New York: Encyclopaedia Britannica,
1929), s.v.; Gregory of Nyssa, "On the Making of Man," in *Nicene and Post-Nicene Fathers*,
volume 5, Gregory of Nyssa, edited by Philip Schaff and Henry Wace (1892; reprinted
Grand Rapids: Wm. B. Eerdmans Publishing Company, 1976), p. 394. See also Duane A.
Priebe, "Viewing the Natural World as Creation: A View from the Early Church," forth-
coming.

[4] Augustine, *Of the City of God* (London, 1610), Book 12, Chapter 10, pp. 499-51. Lu-
dovico Vives's commentary of 1522 on this specifies such sources as Pliny, Aristotle, Plato,
Cicero, Laertius, Mela and Diodorus for the existence of people and other beings before
Adam. See Ernest A. Strathmann, "Ralegh on the Problems of Chronology," *Huntington
Library Quarterly*, 11 (1948): 129-48, on p. 132.

written by the celebrated Jewish philosopher of the Middle Ages, Maimonides, while in Julian Halevi's *The Kuzari*, written between 1130 and 1140 in Spain, allowance was also made for the possibility of the existence of previous worlds.[5]

The stigma of heresy that Augustine had attached to the speculative metaphysics and arcane chronologies of early preadamism proved impossible to stifle even in the face of the more compelling empirical challenges to biblical genealogy that surfaced in the wake of the voyages of reconnaissance. The existence of "monstrous races" had long raised questions for the church fathers, and these only intensified as exploration proceeded apace and various human-like forms made their appearance on didactic *mappaemundi*. Savage pictures, Jonathan Swift remarked, filled up the *terrae incognitae* of geographers' maps. In particular the possible existence of the antipodes was an embarrassment because they implied the presence of a race not descended from Adam. Now, with the "discovery" of America, with all its anthropological and geographical threats to conventional history, some responded by modifying the standard chronicles of creation and accordingly were—as often as not—accused of keeping schools of atheism, peddling skepticism and harboring heretics. Allegations of preadamism were thus among the charges laid at the feet of men like Sir Walter Raleigh (1554-1618) and Thomas Harriot (1560-1621). In Raleigh's case, of course, his theorizing was always conducted within the confines of scriptural authority, but his computational strategy was to seek for the greatest amount of time that the Hebrew text would allow. As for Harriot, it was his experience of exploration in Virginia that, together with his work on biblical chronology, raised questions about the origin of the 'American Indians.' Perhaps it was for these reasons that both Raleigh and Harriot, and indeed Christopher Marlowe, were branded with holding to the preadamite heresy, and belonging to a circle of atheists which impiously and impudently persisted in affirming that American Indian archaeology gave evidence of artifacts that predated Adam by thousands of years.[6] Indeed for chronologists

---

[5] These are discussed in R.H. Popkin, "The Development of Religious Scepticism and the Influence of Isaac la Peyrère's Pre-Adamism and Bible Criticism," in *Classical Influences on European Culture, A.D. 1500-1700*, ed. R.R. Bolgar (Cambridge: Cambridge University Press, 1976), 271-280; Richard H. Popkin, "The Pre-Adamite Theory in the Renaissance," in *Philosophy and Humanism. Renaissance Essays in Honor of Paul Oskar Kristeller* ed. Edward P. Mahoney (Leiden: E.J. Brill, 1976), 50-69.

[6] On early responses to the question of the antipodes and the existence of "monstrous races" see, F.S. Betten, "St. Boniface and the Doctrine of the Antipodes," *American Catholic Quarterly* Review 43 (1918): 654-663; John Block Friedman, *The Monstrous Races in Medieval*

of world history, according to John Dove, the greatest moral problem was that the annals of pagan history seemed to confirm the speculations of those infidels who claimed the existence of "genealogies more ancient than Adam."[7]

To be sure, direct evidence of Raleigh's and Harriot's alleged commitment to Preadamism is lacking, but any inklings in that direction would have been reinforced by the writings of Renaissance scholars like Paracelsus and Giordano Bruno. Paracelsus, for example, was propelled towards polygenism by the sheer presence of newly discovered races. It was, he confessed, difficult to believe that the inhabitants of the "hidden islands" were descended from Adam, and while he was convinced that they had no souls, he suggested that "these people are from a different Adam." As he further explained:

> The children of Adam did not inhabit the whole world. That is why some hidden countries have not been populated by Adam's children, but through another creature, created like men outside of Adam's creation. For God did not intend to leave them empty, but had populated the miraculously hidden countries with other men.

As for Bruno, he was convinced that the Ethiopians, the American Indians, Pygmies, and various species of giants and troglodytes were not "sprung from the generative force of a single progenitor." Rather their origins could be traced back to one of the three sources that comprised his tri-partite cosmological scheme:

> The regions of the heavens are three; three of air; the water
> Is divided into three; the earth is divided into three parts.
> And the three races had three Patriarchs,
> When mother Earth produced animals, first
> Enoch, Leviathan, and the third of which is Adam;
> According to the belief of most of the Jews,
> From whom alone was descended the sacred race.[8]

---

*Art and Thought* (Cambridge: Cambridge University Press, 1981); David Woodward, "Medieval Mappaemundi," in J.B. Harley and David Woodward (eds), *The History of Cartography. Volume One. Cartography in Prehistoric, Ancient, and Medieval Europe and the Mediterranean* (Chicago: University of Chicago Press, 1987), 319, 332. On the skeptical reputations of Raleigh and Harriot, see Ernest A. Strathmann, "The History of the World and Ralegh's Scepticism," *Huntington Library Quarterly*, 3 (1940): 265-287; Pierre Lefranc, *Sir Walter Ralegh. Ecrivain, l'Oeuvre et Les Idées* (Quebec: Les Presses de l'Université Laval, 1968), chapter 12; Jean Jacquot, "Thomas Harriot's Reputation for Impiety," *Notes and Records of the Royal Society* 9 (1952): 164-187.

[7] John Dove, *A Confutation of Atheism* (London, 1677), 176-177. The challenges from ancient Chinese chronology are discussed in Edwin J. Van Kley, "Europe's 'Discovery' of China and the Writing of World History," *American Historical Review*, 76 (1971): 358-385.

[8] These extracts from Paracelsus and Bruno appear in J.S. Slotkin, ed., *Readings in Early Anthropology* (London: Methuen, 1965), 42-43.

The availability of works from the pens of these savants was supplemented by those general surveys of human knowledge that did not hesitate to lay before the reading public the alternative annals of world history passed down from antiquity. Montaigne (1533-1592), for example, probably drew on Ludovico Vives's commentary on Augustine's *City of God* to inform his readers that Herodotus had learned from Egyptian priests of their multi-thousand year history; that according to some the world was eternal; and that

*Cicero* and *Diodorus* said in their daies, that the Chaldeans kept a register of foure hundred thousand and odde yeares. *Aristotle, Plinie,* and others, that Zoroastes lived sixe thousand yeares before *Plato.* And *Plato* saith, that those of the citty of *Sais,* have memories in writing of eight thousand yeares, and that the towne of Athens, was built a thousand yeares before the citty of Sais.[9]

Such figures manifestly problemized the standard six-thousand-year-old creation at the heart of Christian theology. And yet as often as not the rehearsing of these ancient cosmologies with their estimated ages of the world, as in the case of men like Montaigne, Charron, and Thomas Lanquet, was with the express object of refuting them.[10] Thus the encounter with ancient chronologies, together with the anthropological awakening precipitated by the voyages of exploration presented more and more challenges to the received interpretation of the biblical chronicles of creation, and thus encouraged speculation about the existence of non-Adamic peoples.

### The Peyrèrean Formula

These earlier fragmentary speculations about possible preadamic worlds were stitched together in what the leading historian of preadamism in the Renaissance, Richard Popkin, calls "the monumentally heretical doctrine" of Isaac de la Peyrère (1596-1676), a Calvinist of Portuguese Jewish origins from Bordeaux.[11]

---

[9] "An Apologie of Raymond Sebond," in Montaigne's *Essays,* translated by John Florio, 3 vols (London: Dent, 1965), vol. 2, book 2, chap. 12, p. 288.

[10] Pierre Charron, *Of Wisdom* (London, 1612); Thomas Lanquet, *Epitome of Chronicles* (London, 1559). On the idea of "paganism" as providing a vocabulary that linked together certain common characteristics of exotic peoples and of the ancient world, and on the belief that the apologetic arguments that had served the Church Fathers in the late Hellenistic world would work just as well for the contemporary heathen, see Michael T. Ryan, "Assimilating New Worlds in the Sixteenth and Seventeenth Centuries," *Comparative Studies in Society and History,* 23 (1981): 519-538.

[11] Popkin, "Development of Religious Scepticism," 275. See also Richard H. Popkin, "Isaac La Peyrère and the Beginning of Religious Scepticism," in *The History of Skepticism*

Truly the theological edifice that Peyrère erected was composed of many conceptual building blocks. There was, understandably, the pressure exerted by pagan genealogies. As agent of the Prince of Condé in Paris he came into contact with many scholars of the day and had immersed himself in the latest information on exploration and early history. Accordingly *Prae-Adamitae* of 1655 was replete with details of Egyptian, Greek, Babylonian and Chinese histories and how these challenged conventional Christian history.[12] But more important even than this was Peyrère's presentation of the *internal* biblical problems of maintaining the traditional Adamic story.

Peyrère began, as the full title of his work makes plain, with Pauline theology. His concern was to explicate St. Paul's words that "Until the law, sin was in the world; but sin was not imputed, when the law was not." Peyrère found unconvincing standard interpretations which took this text as referring to the Mosaic law, and he proceeded to argue his case for it as law given to Adam. His conviction was that the history of ceremonial Judaism must be traced back beyond Moses to Adam himself. As he put it:

Long before Moses there were other Ordinances prescribed, and commended to the Jews, other Ceremonies instituted, other Laws of God decreed and confirmed for that holy and elected People. And in this place, I mean the Jews, not onely [sic] the Sons of Abraham who are called the Seed of Abraham, but also the fore-fathers of Abraham, the Posterity of Adam.

The consequences of this hermeneutic move were immediate: there must have been human beings on earth before Adam. As he summarized his thesis:

For that law was either to be understood of the Law given to Moses or of the law given to *Adam* . . . if that law were understood of the law given to *Adam*, it must be held that sin was in the world before *Adam* and until *Adam* but that sin was not imputed before *Adam*; Therefore other men were to be allowed before *Adam* who had indeed sinn'd, but without imputation; because before the law sins wer [sic] not imputed.[13]

Taken together, the evidence of new geographical and anthropological data combined with internal biblical exegesis, helped lay

*from Erasmus to Spinoza* (Berkeley: University of California Press, 1969); idem, *Isaac la Peyrère (1596-1676): His Life, Work and Influence* (Leiden: Brill, 1987); idem, "The Marrano Theology of Isaac La Peyrère," *Studi Internazionali di Filosofia*, 5 (1973): 97-126.

[12] Peyrère was an expert on the Eskimos and was able to exploit his knowledge of Inuit culture and history to fortify his Preadamite case. He was the author of *Relation du Groenland* (Paris, 1647) and *Relation de l'Islande* (Paris, 1663). The former was translated and edited by Adam White and published by the Hakluyt Society in 1855.

[13] Isaac de la Peyrère, *Men Before Adam. Or a Discourse upon the Twelfth, Thirteenth, and Fourteenth Verses of the Fifth Chapter of the Epistle of the Apostle Paul to the Romans. By Which are Prov'd, That Men Were Created before Adam* (London, 1656), 4, 19.

the foundation of biblical criticism.[14] Extrabiblical data simply had to be accorded a role in biblical hermeneutics. But Peyrère's project did not stop even here. To appreciate fully its conceptual mainsprings, his Messianic vision must be taken into account. For the objective of his preadamite speculations that sets it apart from earlier inklings of the theory was his concern to separate out the Jewish experience from the rest of world history. The theological fulcrum of sacred history was, as he saw it, the centrality of Judaism, and the Preadamite theory was designed to cut a deep gorge between the Jews and the Gentiles, namely, between the history of the Adamites and that of the Preadamites. Indeed Peyrère had already published a pro-Semitic work, *Du Rappel des Juifs* in which he elaborated his vision of world history. Its thesis was that the Bible deals only with the Jews who play the leading role in Providential history. The Gentiles or, better, preadamites merely look on as the divine drama is played out in the world, although they participate in the benefits of God's election of his own people. The political implications of this scenario for Peyrère's France were immediate: France should admit Jews in order to hasten the nation's conversion. All in all *Du Rappel des Juifs* represented an effort to hold in balance the Bible, scientific knowledge, concern for the Jews, and French nationalism.[15]

Taken in its full context, then, Peyrère's preadamite theory was fundamentally a *theological* project, universalistic in impulse and Messianic in character. And yet by proposing the altogether simple idea that people had existed for millennia before Adam, Peyrère was able to reconcile the shortness of biblical chronology with the latest findings of geography, anthropology, and archaeology. The polygenetic (though entirely humanitarian) thrust of his scheme accordingly earned for him an acknowledged place in the early history of the human sciences.[16] Besides, his theory had

---

[14] See Richard H. Popkin, "Biblical Criticism and Social Science," *Boston Studies in the Philosophy of Science*, 14 (1974): 339-360.

[15] See Myriam Yardeni, "La Religion de La Peyrère et "Le Rappel des Juifs'," *Revue d'Histoire et de Philosophie Religieuse*, 51 (1971): 245-259. Such speculations have led several scholars to insist that Peyrère was a Marrano, but conclusive evidence for this is apparently lacking. See Anthony Grafton, "A Vision of the Past and Future," *Times Literary Supplement*, 1988, February 12-18: 151-152.

[16] See, for example, A.C. Haddon, *History of Anthropology* (London: Watts & Co., 1910), 52; D.R. McKee, "Isaac de la Peyrère, A Precursor of Eighteenth-Century Critical Deists," *Publications of the Modern Language Association*, 56 (1944): 456-485; T.K. Penniman, *A Hundred Years of Anthropology* (London: Duckworth, second edition 1952), 53; Juan Comas, *Manual of Physical Anthropology* (Springfield, Ill.: Charles C. Thomas, 1960), 77, 80; D.C. Allen, *The Legend of Noah: Renaissance Rationalism in Art, Science, and Letters* (Urbana, Ill.: University of Illinois Press, 1963), 137; Thomas F. Gossett, *Race: The History of an Idea in America* (Dallas: Southern Methodist University Press, 1963), 15; James Johnson, "Chronological Writing:

more general biogeographical implications. Some of these were seized upon by a Dutch naturalist, Abraham van der Myl, who used it to argue that the entire fauna of the New World, animal as well as human, had been separately created.[17] Then there was Peyrère's suggestion that the Mosaic flood was not universal. Even critics of preadamism could find this attractive, as in the case of Bishop Edward Stillingfleet (1635–1699) who, while repudiating the suggestion of the existence of men and women before Adam as fiercely as England's Lord Chief Justice Matthew Hale (1609-1676), nevertheless found the idea of a local flood distinctly appealing. For him the deluge was demographically, not geographically, universal.[18]

However Peyrère's arch-heresy was to be faced, things could never be the same afterwards. It now became difficult to maintain that the Bible was the sole authoritative account of human origins, and this suspicion, taken with Peyrère's own doubts as to the Mosaic authorship of the Pentateuch and his observations on the unreliability of the *textus receptus*, served to keep the preadamite theory firmly tied to skeptical moorings. No doubt it was for this reason that Peyrère's preadamism found its way into Lecky's *History of the Rise and Influence of the Spirit of Rationalism in Europe* as a sort of half-way house towards free thought.[19]

Whatever his own hopes for the fate of his preadamite theory,

Its Concept and Development," *History and Theory*, 2 (1962): 124-145; Margaret T. Hodgen, *Early Anthropology in the Sixteenth and Seventeenth Centuries* (Philadelphia: University of Pennsylvania Press, 1964), 272-276; Marvin Harris, *The Rise of Anthropological Theory: A History of Theories of Culture* (New York: Thomas Y. Crowell, 1968), 89; Fred W. Voget, *A History of Ethnology* (New York: Holt, Rinehart and Winston, 1975), 58-59; Nancy Stepan, *The Idea of Race in Science: Great Britain 1800-1960* (London: Macmillan, 1982), 29; Donald K. Grayson, *The Establishment of Human Antiquity* (New York: Academic Press, 1983), 140-142; Paolo Rossi, *The Dark Abyss of Time: The History of the Earth and the History of Nations from Hooke to Vico*, translated by Lydia G. Cochrane (Chicago: University of Chicago Press, 1984), 132-136; Bruce G. Trigger, *A History of Archaeological Thought* (Cambridge: Cambridge University Press, 1989), 112.

[17] Abraham van der Myl, *De Origine Animalium et Migratione Populorum* (Geneva, 1667). More generally on this influence see, Janet Browne, *The Secular Ark. Studies in the History of Biogeography* (New Haven and London: Yale University Press, 1983), 12-15.

[18] See Matthew Hale, *The Primitive Origination of Mankind Considered and Examined According to the Light of Nature* (London, 1677); Edward Stillingfleet, *Origines Sacrae, or a Rational Account of the Grounds of Christian Faith* (London, 1662). On Stillingfleet see Richard H. Popkin, "The Philosophy of Bishop Stillingfleet," *Journal of the History of Philosophy*, 9 (1971): 303-319; Robert T. Carroll, *The Common-Sense Philosophy of Religion of Bishop Edward Stillingfleet* (The Hague: Martinus Nijhoff, 1975).

[19] William Edward Hartpole Lecky, *History of the Rise and Influence of the Spirit of Rationalism in Europe*, 2 vols. (London: Watts & Co., 1910, first pub. 1866), vol. 1, pp. 107-108. Lecky pointed out that Peyrère rejected the Mosaic authorship of the Pentateuch, radically reinterpreted the doctrine of original sin, and restricted the extent of the miraculous, thereby establishing the foundations of biblical criticism. See also Andrew D. White, *A History of the Warfare of Science with Theology in Christendom* (New York: George Braziller, 1955, orig. 1894), vol. 1, p. 255.

Peyrère found his proposals pilloried by the president and council of Holland and Zeeland in November 1655, and by the Bishop of Namur a month later on Christmas Day. Within a few weeks Peyrère was arrested in Brussels, imprisoned and taken to Rome where, within the year, he had recanted before Pope Alexander VII in order to obtain release. But Peyrère so structured his official recantation that he never explicitly admitted that Preadamism was false, and while living out his last days in a seminary outside Paris, he continued to accumulate further evidence to support his grand profanity.

Although the theory attracted no more than the merest handful of supporters in the years that followed, refutation after refutation was issued in a desperate effort to exorcise the religious mind of the ghost of the preadamite. Indeed Popkin has been able to specify more than two dozen antidotes to Peyrère's preadamism before the end of the seventeenth century, perhaps the most important of which was that by Samuel Desmarets which was later repeated by Diderot. And he tells a similar story for much of the eighteenth century, too. Moreover when the preadamites did crystallize they were invariably to be found in the bad company of skeptics or esoterics. As for the former, Popkin presents evidence that preadamism was adopted by Voltaire, while on the latter score no better candidate may be identified than the Irish parliamentarian Francis Dobbs who is reported to have supposed that there was a non-Adamic human race conceived by an intrigue of Eve and the Devil, though in fact his views were more similar to those of Nathaniel Brassy Halhed whose exposure to the Hindu scriptures brought him to the horrible brink of preadamism.[20]

Whatever its taint of heresy, there were, nonetheless, those whose rejection of polygenism was altogether hesitant or who set about modifying the basic Peyrèrean formula to suit their own purposes. Consider the case of Nathaniel Lardner (1684-1768), a non-conformist scholar whose theological convictions developed from Baxterian Calvinism through Arminianism to a modified Arianism. His greatest achievements were his *Credibility of the Gospel History* published between 1727 and 1757 and *A Large Collection of Ancient Jewish and Heathen Testimonies to the Truth of the Christian Religion*, a four-volume treatise issued between 1764 and

---

[20] Popkin, *Peyrère*, 81, 121, 131, 133. Dobbs's strange millennial politics are to be found in *Memoirs of Francis Dobbs, Esq. Also Genuine Reports of His Speeches in Parliament on the Subject of an Union, and His Prediction of the Second Coming of the Messiah; With Extracts from his Poem on the Millennium* (Dublin, Jones, 1800). I am grateful to Dr. Myrtle Hill for drawing my attention to this work.

1767; in these he strove to defend the authenticity of the Bible against deistic critics.[21] To be sure, Lardner's account of the Genesis narrative was self-confessedly literalistic; but in an essay he penned in 1753 on the subject of the creation and the fall, he admitted that there were "not a few difficulties in the account, which Moses has given of the world, and of the formation, and temptation, and fall of our first parents." The greatest of these centered on human origins, and Lardner conceded that the belief that "all mankind have proceeded from one pair"—the sheet anchor of the Christian conception of creation—could not be absolutely established from contemporary empirical evidence without the bolstering support of scripture. Indeed he did confess that "many pairs, resembling each other, might have been formed by God, the Creator, at once, in several, and remote countries, that the earth might be peopled thereby." But he himself felt constrained to reject this possibility, diffidently adding that "the account of Moses, I suppose, may be relied upon." As to the source of racial differentiation, Lardner was forced to resort to the "difference of climates, with the varieties of air, earth, water" as the means of making "sensible alterations and differences in one and the same species." Moreover, Lardner did not hesitate to point to what he saw as the moral implications of the Mosaic monogenetic account, namely, that common descent from the same parents should "abate exorbitant pride" and re-affirm the universal brotherhood of the human race.[22]

For those unnerved by charges of heresy, the option of modifying the basic scheme seemed one possible means of retaining its explanatory power while deflecting its heterodoxy. Thus the suggestion of co-adamites began to receive an airing. The idea here was that God had created several humans in different locations at the same time as Adam—a view which clearly removed the stigma attached to the notion that the world was inhabited before Adam. This version was anonymously promulgated in 1732 in a tract entitled *Co-Adamitae, or an Essay to Prove the Two Following Paradoxes, viz. I. That There Were Other Men Created at the Same time with Adam, and II. That the Angels did not Fall.* The author's main objective was to relieve internal tensions in the biblical narrative by answering exegetical questions like where Cain got his wife

---

[21] Biographical details are available in *Dictionary of National Biography*, s.v., and in Ian Sellers, "Lardner, Nathaniel," in *The New International Dictionary of the Christian Church*, ed., J.D. Douglas (Exeter: Paternoster, 1974), q.v.

[22] Nathaniel Lardner, "An Essay on the Mosaic Account of the Creation and Fall of Man" (first published in 1753), in *The Works of Nathaniel Lardner*, D.D. 11 vols (London, 1788), vol. 11, pp. 227-251, on pp. 227, 244-245.

and who peopled his city. But subsequently, as we shall shortly see, this self-same system was advanced with rather more scientific objectives in mind.

It is time to pause and take stock of preadamism's fortunes thus far. While the context within which the preadamite was talked about was almost invariably theological, it would clearly be mistaken to think of that vocabulary as hermetically sealed off from other fields of discourse. Indeed it was not so much that there were a range of demarcated conversations—geographical, anthropological, theological, and so on—between which there was cross-communication. Rather, geographical and anthropological considerations just were part of the theological frame of reference, and vice versa. Of course this does not mean that there was universal agreement among the conversationalists. And as we have seen, advocates of preadamism, either by self appointment or public accusation, frequently found themselves marginalized on the skeptical fringes of the interchange. All this, however, would soon change as preadamite vocabulary, in one form or another, became more common in the emerging language of the human sciences.

## II. Preadamism and Racial Science

By the end of the eighteenth century pre-adamite and co-adamite relics began to be reported, at first sporadically, but soon with increasing frequency. The setting in which they were now to be found however was beginning to alter. Bit by bit the typically exegetical anxieties of the biblical expositor were supplemented and, in many instances, replaced by the newer challenges that emerged from anthropology, archaeology, geology, and later linguistics. To be sure, there was no sudden or radical shift from theology to science as the conceptual mainstay of preadamism; rather the balance of debate merely moved from intramural biblical hermeneutics to "scientific" questions about the origin and unity of the human race. Correlated with this cognitive transfer, moreover, was a shift to something of a more exclusively English-language context. This is not to say that advocates of preadamism were restricted solely to the English-speaking world; but it is to note that it was in Britain and America that supporters of the system were most clearly discernible. No doubt the reason for this conceptual migration had to do with the continuing vitality of natural theology in eighteenth and nineteenth century Britain and America as the common context for science. Compared with France, for example, where men of science joined in the attack against Christianity because they saw religion bolstering a social order that they themselves rejected, in Britain and America practitioners of science were generally drawn from the ranks of the religious and social establishment. Since British and American students of nature and human nature thus continued to ground their scientific endeavors in natural theology, frequently in a biblicist mold, it was necessary to find some means of integrating the findings of the new sciences of ethnology, archaeology, human geography and anthropology with theological principles. In this context preadamism therefore provided one convenient—albeit frequently contested—vehicle by which human science and biblical theology could remain bound together. In continental Europe, as we shall presently see, preadamism where it survived remained rather more exclusively a *theological* option re-emerging in the late nineteenth century and persisting right into the twentieth century among those conser-

vatives for whom the idea of Adam retained conceptual force. But
first we turn to the ways in which the notion of multiple human
creations could be put to work for the purposes of creating a
science of race.

## Preadamism, Polygenism and Race

This chapter reveals at least two significant departures from
earlier preadamism. First, the theory's newest advocates—who
frequently, but not invariably, connected it up to polygenism—
commonly deployed the scheme in the cause of racial apologetic,
namely as a justification for white superiority. Preadamism in this
mode thus broke with the egalitarian universalism of the original
Peyrèrean formula according to which all men and women, irre-
spective of racial background or religious belief, automatically
enjoyed the benefits of Israel's redemption. Late eighteenth and
nineteenth century preadamism, however, could do more than
serve as a legitimating tool for the racial ideologue; it could be put
to scientific work. That is to say, beliefs about preadamism,
whether supportive or antagonistic, could be used to establish
what might be called explanatory alignments. Its champions, for
example, generally found themselves committed to the notion of
fixity of type—the original racial differences established at the
creation of the separate human groups were regarded as persis-
tent. By contrast those who retained a monogenetic belief in universal descent from an Edenic pair had little option but to account
for racial diversity by appealing to the modifying effects of environment, diet, or innate tendency. Thus, advocates of Preadamism generally opposed the acclimatization theory, while de-
tractors ordinarily welcomed it; the former commonly held to
fixity of type, the latter generally tended towards some version of
developmentalism.[1] Conceptual maneuvers of this kind suggest
that here at least metaphors of warfare or divorce only serve to
obscure our understanding of the historical relationship between
science and religion.

The more specifically scientific context for preadamite discourse towards the end of the eighteenth century is perhaps nowhere more evident than in the polygenetic account of human

---

[1] On this topic see David N. Livingstone, "Human Acclimatization: Perspectives on a
Contested Field of Inquiry in Science, Medicine and Geography," *History of Science*, 25
(1987): 359-394. See also Michael Osborne, "The Société Zoologique d'Acclimatation and
the New French Empire: The Science and Political Economy of Economic Zoology during
the Second Empire" (Ph.D. thesis, University of Wisconsin-Madison, 1987), especially
chapter 5 on "The Scientific Basis of Acclimatization in France."

origins advanced by the Scottish student of anthropology, Lord Kames (1696-1782). It was in his *Sketches of the History of Man* first published in 1774 during that spectacular period of cultural life in Scotland known as the Scottish Enlightenment, that Kames denied the organic unity of the human race. Human differences, he felt, were just too great to believe in the indivisibility of the species. Like other key figures in the Scottish Enlightenment, notably Thomas Reid, Dugald Stewart, and Francis Hutcheson, Kames sought for a coherent theory of the human constitution. Although Scottish Enlightenment philosophy generally lodged its conception of the universality of human nature in an assumed monogenism, data accumulated by geographers and anthropologists brought Kames to the brink of polygenetic coadamism as he toyed with the idea that God had created in different climatic regimes many human pairs with appropriately regulated physiologies. Buffon's belief that Africans, Asians, Native Americans, and Europeans, were the same human species tinged only with the colors of climate, Kames found unconvincing. If climate was so determinative, he wondered, then why were diverse racial groups to be found in similar climatic regimes, and why did North America's regional climates not produce corresponding racial types? Not surprisingly Kames was no more impressed with Montesquieu's tracing the attributes of mind to a causal environment. Thus if climate could not be called upon to explain racial diversity, and migration likewise failed to account for the existence of American Indians, polygenism seemed the only option. The Native Americans, it appeared, were a separate post-Adamic creation. Climate, accordingly, did not produce human variety; human varieties were produced for particular climates. When transported elsewhere races inevitably degenerated. And yet for all the evidence he had accumulated, Kames ultimately drew back from this skeptical heresy and, in a desperate effort to avoid the charge of infidelity, he proposed that God had impressed upon the human race an immediate change of constitution at the time of the Tower of Babel to fit it for its diaspora.[2]

Still, for all his apparent tentativeness Kames succeeded in establishing polygenism as a viable scientific account of racial de-

---

[2] Henry Home [Lord Kames], *Sketches of the History of Man*, 2 vols. (Edinburgh, 1774). See the discussion of Kames in William Stanton, *The Leopard's Spots. Scientific Attitudes toward Race in America, 1815-59* (Chicago: University of Chicago Press, 1960), 15-16; George Stocking, *Race, Culture, and Evolution. Essays in the History of Anthropology* (Chicago: University of Chicago Press, 1982), 44-45; Robert Wokler, "Apes and Races in the Scottish Enlightenment: Monboddo and Kames on the Nature of Man," in Peter Jones (ed.), *Philosophy and Science in the Scottish Enlightenment* (Edinburgh: John Donald Publishers, 1988).

velopment, and his views were promulgated by those less unnerved by its taint of profanity. The English surgeon Charles
White (1728-1813), to take just one instance, threw his weight
behind the proposal. He had had the opportunity of poring over
the cabinet of human skulls in the possession of John Hunter, and
turned to the multiple creation hypothesis as a means of explain-
ing the apparent gradations in human and animal anatomy. In his
*Account of the Regular Gradation in Man* (1799) he thus asserted that
"the various species of men were originally created and sepa-
rated, by marks sufficiently discriminative."[3]

The driving force behind Kames's suspicions of multiple Ad-
amism, of course, was somewhat different in emphasis from that
of Peyrère's polygenic Preadamism; namely, it was rather more
scientific than theological. But there were nonetheless those anxious about the more broadly theological implications of such con-
cessions, and their defense of the traditional Adamic narrative
committed them to an alternative account of racial differentiation.
Perhaps the most conspicuous critic of Kames from this perspective was Samuel Stanhope Smith (1751-1819), clergyman-profes-
sor of moral philosophy at the College of New Jersey (Princeton),
and subsequently its president. In his *Essay on the Causes of the
Variety of Complexion and Figure in the Human Species* which first
appeared in 1787 and then in an expanded form in 1810, Smith
determined to show that the unity of the human species was not
only good theology; it was good science, and even more important, good moral philosophy. The ironies in the debate between
Smith and Kames are manifold. For in Smith, the self-appointed
defender of orthodoxy, we find one who *rejected* special provi-
dences and thereby *naturalized* explanations by conceiving of God
as building certain intrinsic qualities into the very stuff of reality.
Kames, the supposed infidel, however, resorted to miracle to
explain the diversification of languages at Babel, and clearly was
far more supernaturalistic in his belief in separate human creations. Smith's achievement in branding Kames as impious re-

---

[3] Charles White, *An Account of the Regular Gradation in Man, and in Different Animals and
Vegetables; and from the Former to the Latter* (London, 1799), 125; for biographical details, see
Charles W. Sutton, "White, Charles" in *Dictionary of National Biography*, s.v. According to
Philip Curtin, the co-adamite theory was specifically applied to Africans by a naval sur-
geon, John Atkins, in 1734. See Philip D. Curtin, *The Image of Africa: British Ideas and Action,
1780-1850* (London: Macmillan, 1965), 41. As early as 1680, Morgan Godwyn—author of
*Negro's & Indians Advocate*—was aware that the "Pre-Adamite whimsey" was being har-
nessed to subserve racist ends. See Richard H. Popkin, "Pre-Adamism in 19th Century
American Thought: 'Speculative Biology' and Racism," *Philosophia*, 8 (1978): 205-239, on p.
212.

quired, as Mark Noll has put it, "considerable rhetorical finesse."[4] In any case, the key issue for Smith in the whole matter was the management of the moral economy, for he believed no "general principles of conduct, or religion, or even of civil policy, could be derived from natures originally and essentially different."[5] Scientific questions about human origins were thus directly implicated in matters of social polity. For Smith what was at stake in researches into historical anthropology was the constitution of society and the maintenance of the social order. Accordingly he drew on Blumenbach's work to minimize the differences between human groups and offered an environmentalist explanation of racial origin,[6] and thus in Smith's work the umbilical links between monogenism and acclimatization emerge clearly—an alliance that appears again and again throughout the nineteenth century.[7]

The writings of James Cowles Prichard illustrate this association, as he progressively moved toward environmentalism in order to preserve his monogenist stance in the face of ever more assertive polygenetic challenges which were, at the same time, staunchly hereditarian. Having grown up as a Quaker and later influenced by evangelicalism, he too was profoundly committed to monogenism and thus found himself looking more and more to climate to explain somatic differences among racial groups. Early in his career he had considered civilization as a race-forming factor arguing that it stimulated variation in a way rather akin to the

---

[4] Mark A. Noll, *Princeton and the Republic, 1768-1822. The Search for a Christian Enlightenment in the Era of Samuel Stanhope Smith* (Princeton: Princeton University Press, 1989), 121.

[5] Samuel Stanhope Smith, *An Essay on the Causes of the Variety of Complexion and Figure in the Human Species* (Cambridge, Mass.: Harvard University Press, 1965; facsimile reprint of 1810 second edition; first pub. 1787), 149. See also the discussion in Herbert W. Schneider, *A History of American Philosophy* (New York: Columbia University Press, 1946), 343-65. The links between the 'natural history of man' and moral philosophy, particularly in regard to the constitution of the human mind, were also crucially important in the Scottish Enlightenment. See P.B. Wood, "The Natural History of Man in the Scottish Enlightenment," *History of Science*, 27 (1989): 89-123.

[6] It should perhaps be noted that Blumenbach believed that there had been "a whole organized preadamite creation [that] has disappeared from the face of our planet." Thus, *The Anthropological Treatises of Johann Friedrich Blumenbach*, translated and edited by Thomas Bendyshe (London: Longman, Green, Longman, Roberts, & Green, 1865), 285.

[7] Such an alliance was not universal. Believers in the standard Adamic history were not invariably committed to the climatic thesis. Samuel Latham Mitchill (1764-1831), for example, although a firm believer in the unity of the human species, felt that climate was insufficient to account for the differences between "tawny man," "black man," and "white man." His suggestion was that there was some internal impulse towards somatic differentiation, a force that he dubbed "the generative influence." Samuel L. Mitchill, "The Original Inhabitants of America Shown to be the Same Family and Lineage with Those of Asia, by a Process of Reasoning not Hitherto Advanced," *Archaeologia Americana: Transactions and Collections of the American Antiquarian Society*, 1 (1820): 225-232. See the discussion in Stanton, *Leopard's Spots*, 9.

processes of domestication. But he increasingly moved towards identifying a correlation between climate and physical type. By thus resorting to the influence of natural environment the door was opened to the drawing of analogies between human and animal development.[8] If the climatic adaptations of animal species could be interpreted naturalistically, there was little reason to exempt the human species from the self-same agents of change. Indeed Prichard's, and for that matter Smith's, racial accounts were inherently "evolutionary" rather than "creationist" and thereby illustrate the alliance between orthodox theology and developmentalist anthropology in the period. For Prichard, however, one component of his anthropological system brought disquiet to many a respectable soul. Deeply convinced of progressive development from savagery toward civilization (although, taken overall, as Stocking makes clear, his thinking is possibly better typified as diffusionist rather than developmentalist), and what seemed to him its self-evident corollary—progress from black to white, Prichard suggested that Adam must have been black.[9]

Smith and Prichard were not alone in their championing of traditional monogenism. On the contrary monogenist partisans were forever telling their readers that theirs was the view of Blumenbach, Buffon, Cuvier, Lamarck, and Montesquieu, and that these savants had vindicated their environmentalist explanations. Besides, numerous lesser witnesses could also be called upon to provide supporting testimony. In 1775 James Adair urged that the American Indians were of Jewish origin, and claimed thereby to

---

[8] Thus Prichard discusses animal acclimatization in *The Natural History of Man; Comprising Inquiries into the Modifying Influence of Physical and Moral Agencies on the Different Tribes of the Human Family* (London: Hippolyte Billiere, 3rd ed., 1848), 33, 37, 60, 66. Accordingly he cited the opinion of M. Poulin in support of the view that "acclimatisation . . . consists in certain permanent changes produced in the constitution of animals, which bring it into a state of adaptation to the climate." ibid., 39. And while Poulin and Prichard stopped short of species transformation, the Lamarckian undergirding of their views on heredity transmission is plain in Prichard's reporting of the following "facts" established by Poulin: "Permanent changes or modifications in the functions of animal life, may be effected by long-continued changes in the habitudes which influence these functions. . . .Hereditary instincts may be formed, some animals transmitting to their offspring acquired habits." ibid., 40. Certainly Prichard never wholeheartedly embraced Lamarckism, but as Stocking points out, he increasingly retreated, like Darwin, towards it. See George W. Stocking, "From Chronology to Ethnology: James Cowles Prichard and British Anthropology, 1800-1850," in George W. Stocking (ed.), *Researches into the Physical History of Man, James Cowles Prichard* (Chicago: University of Chicago Press, 1973), lxxx; Stepan, *Idea of Race* (cit. n. 19), 31-40

[9] See George W. Stocking Jr., *Victorian Anthropology* (New York and London: The Free Press, 1987), 48-53. Later Prichard drew back from the heterodoxy of making Adam black, though this abandonment of primitive blackness was probably less to do with social niceties as to changes in his attitude to the theories of Jacob Bryant who regarded early culture-bearers as descendants of Ham.

have overthrown Kames's entire polygenist system, and more particularly the "wild notion" that the North American Indians were "Prae-Adamites."[10] Again in 1816 James Haines M'Culloh, a medical practitioner, just assumed that the foremost naturalists had arrived at a consensus on the unity of the human species—a view endorsed by Philadelphia paleontologist Richard Harlan in 1822—and went on to show to his own satisfaction that the Indian could have arrived by migration to America from the original locus of creation.[11] As works of this type became available, apologists for the traditional Mosaic cosmogony could confidently reaffirm the conventional Adamic history and report that science confirmed either Usher's 4004 B.C. date or the Septuagint's alternative 5872 B.C.[12]

The doctrine of the unity of the human species that undergirded the monogenism of Smith and Prichard came to the surface no less in theological works than in scientific discourse. In Smith's own institution, Princeton, the Presbyterian tradition steadfastly hewed to the monogenist line. In 1829, for example, Archibald Alexander (1772-1851), who drew up the blue-print that would guide Princeton theology for a century and through whom Old School Presbyterianism received its baptism in Scottish Common Sense philosophy, turned his attention to the relationship between scripture and nature. Here he dismissed the idea "that there were men before Adam" as "destitute of all shadow of proof," suggested that the varieties of the human species were due to "the great difference of climate and other circumstances of the nations of the earth," and declared as a "prejudice without foundation [the belief] that the colour of the whites

[10] James Adair, *The History of the American Indians* (London, 1775), 11. The same viewpoint comes through in William Robertson, *The History of America*, 4 vols. (Edinburgh: Stirling & Slade, 1819), vol. 2, pp. 25-49.

[11] J.H. M'Culloh, *Researches on America: Being an Attempt to Settle Some Points Relative to the Aborigines of America &c* (Baltimore, 2nd ed., 1817), xi-xii; Richard Harlan, "Remarks on the Variety of Complexion, and National Peculiarity of Feature," in *Medical and Physical Researches; or, Original Memoirs in Medicine, Surgery, Physiology, Geology, Zoology, and Comparative Anatomy* (Philadelphia, 1835), 521-589. See also the discussion of M'Culloh and Harlan in Stanton, *Leopard's Spots*, 10-11. Later the same strategy was apparent in J.L. Cabell, *The Testimony of Modern Science to the Unity of Mankind; Being a Summary of the Conclusion Announced by the Highest Authorities in the Several Departments of Physiology, Zoölogy, and Comparative Philology in Favor of the Specific Unity and Common Origin of all the Varieties of Man* (New York: Robert Carter & Brothers, 1859). Cabell was a professor of Comparative Anatomy and Physiology in the University of Virginia.

[12] A typical work of this genre was Thomas Wood's *The Mosaic History of the Creation of the World: Illustrated by Discoveries and Experiments Derived from the Present Enlightened State of Science: To Which is Prefixed, the Cosmogony of the Ancients: With Reflections, Intended to Promote Vital and Practical Religion* (London, 2nd ed., 1818).

was that of the first man."[13] Later, at Princeton Theological Seminary, Charles Hodge similarly defended the unity of the human race as did Benjamin B. Warfield who, as we shall see, was able to integrate it both with evolutionary theory and with a newer form of preadamism.

However dismissive its critics, whether on scientific or theological grounds, the preadamite theory began to attract an ever growing number of defenders during the nineteenth century. In 1800, Edward King, a Fellow of the Royal Society, called attention to the "many proofs and arguments that may be derived from the Holy Scriptures themselves, which tend to show . . . that *the commonly received opinion, that all mankind are the sons of Adam . . .* is directly contrary to what is contained there." To him the Genesis narrative depicted two quite different creation stories, the first dealing with mankind in general, the second with an individual, Adam. By this exegetical move, King could at once explain troublesome details in the biblical text, account for the racial diversity of the world, and provide his readers with a racial ideology tailor-made to serve the interests of white superiority.[14] Not that King was the first to promulgate such a view. Earlier, in 1774, Edward Long, a former Jamaican judge had taken the same stand, urging that the highest level of Orang-Outang shaded almost imperceptibly into the Negro, and bolstering his polygenetic racism by appealing to the authority of David Hume who had grave suspicions that the universality of human nature would be fractured unless non-white races were regarded as having originally distinct constitutions.[15]

But now as the groundswell of polygenism gathered force during the first half of the new century, particularly in the United

---

[13] Archibald Alexander, "The Bible, A Key to the Phenomena of the Natural World," *Biblical Repertory and Princeton Review*, 1 (1829): 101-120. Reprinted in Mark A. Noll (ed.), *The Princeton Theology. Scripture, Science, and Theological Method from Archibald Alexander to Benjamin Warfield* (Grand Rapids: Baker, 1983), 92-104, on pp. 103, 100, 104. Alexander speculated—like others after him, notably Alexander Winchell and Hugh Miller—that Adam was neither white nor black. His suggestion was that it was "much more probable that our first parents were red men or of an olive or copper colour." Alexander based this judgment on the fact that "the radical signification of Adam is red." ibid., 104. See also Charles Hodge, "The Unity of Mankind," *Biblical Repertory and Princeton Review*, 31 (1859): 103-149.

[14] Edward King, "Dissertation Concerning the Creation of Man," in *Morsels of Criticism, Tending to Illustrate Some Few Passages in the Holy Scriptures upon Philosophical Principles and an Enlarged View of Things* (London, 1800), 70-71.

[15] Edward Long, *History of Jamaica* (3 volumes, London, 1774). David Hume's polygenetic racism is referred to in Harris, *The Rise of Anthropological Theory*, op. cit., 88. It has been argued that too much has been made of Hume's racial sentiments, coming as they do from a lengthy footnote to his 1754 essay "Of National Characters." See P.J. Marshall and Glyndwr Williams, *The Great Map of Mankind. British Perceptions of the World in the Age of Enlightenment* (London: J.M. Dent, 1982), 246

States, under the influence of men like Samuel G. Morton, Josiah Nott, George R. Gliddon, Charles Pickering, and Louis Agassiz, its value as racial apologetic became more apparent than ever.[16] The careers of these men and their polygenetic vision have been scrutinized elsewhere and need not be detailed here. Suffice it to say that the motivation of these individuals was thoroughly anthropological and archaeological, rather than theological. Their concern was to account for the anthropometric range and ethnological patterns displayed among the human races rather than to square biblical chronology with the latest science, and accordingly many of them drifted into a kind of secular preadamism. This certainly did not mean that their "science" was devoid of ideological bias. The story of Morton's "finagling" of anthropometric data to bring out the supposed inferiority of certain racial groups has been engagingly told by Stephen Jay Gould. And such Mortonite practices were perpetuated by his lesser-known successor at the Philadelphia Academy of Sciences, J. Aitken Meigs. Meigs further manipulated Morton's data to get an even larger cranial capacity for the Teutons, to decrease the measurement for certain Native American groups, and to bring down the average score for Blacks. All of this was to bolster the ideology of polygenism that reigned within this circle of American anthropologists.[17]

In Britain broadly similar conceptual alignments are apparent in the writings of Lieutenant-Colonel Charles Hamilton Smith (1776-1859). Smith was both a soldier and an amateur naturalist

---

[16] The writings of these and many other individuals were popularized in the unsubtly titled work by John Campbell, *Negro-mania: being an Examination of the Falsely Assumed Equality of the Various Races of Men; Demonstrated by the Investigations of Champollion, Wilkinson, Rosellini, Van-Amringe, Gliddon, Young, Morton, Knox, Lawrence, Gen. J.H. Hammond, Murray, Smith, W. Gilmore, Simms, English, Conrad, Elder, Prichard, Blumenbach, Cuvier, Brown, Le Vaillant, Carlyle, Cardinal Wiseman, Burckhardt, and Jefferson. Together with a concluding chapter, presenting a comparative statement of the condition of the Negroes in the West Indies before and since emancipation* (Philadelphia: Campbell & Power, 1851).

[17] I have re-examined Meigs's data and compared them with Morton's. While Morton used three Iroquois skulls, Meigs dropped the sample size to two; and whereas Morton used 18 German skulls, Meigs reduced the number to 15 but was still able to get an average cranial size for Teutons that was 5 cubic inches greater than Morton's figure. See J. Aitken Meigs, "The Cranial Characteristics of the Races of Men," in J.C. Nott and George R. Gliddon, *Indigenous Races of the Earth* (Philadelphia, 1857), 203-352. Relevant material on American polygenism more generally may be found in the following works: Thomas F. Gossett, *Race: The History of an Idea in America* (Dallas: Southern Methodist University Press, 1963); Stephen Jay Gould, *The Mismeasure of Man* (New York: Norton, 1981); Reginald Horsman, *Josiah Nott of Mobile: Southerner, Physician, and Racial Theorist* (Baton Rouge: Louisiana State University Press, 1987); Edward Lurie, *Louis Agassiz. A Life in Science* (Chicago: University of Chicago Press, 1960); David N. Livingstone, *Nathaniel Southgate Shaler and the Culture of American Science* (Tuscaloosa and London: University of Alabama Press, 1987); Stanton, *Leopard's Spots*, op. cit.; Popkin, "Pre-Adamism in 19th Century Thought," op. cit.; Robert E. Bieder, *Science Encounters the Indian, 1820-1880. The Early Years of American Ethnology* (Norman: University of Oklahoma Press, 1986).

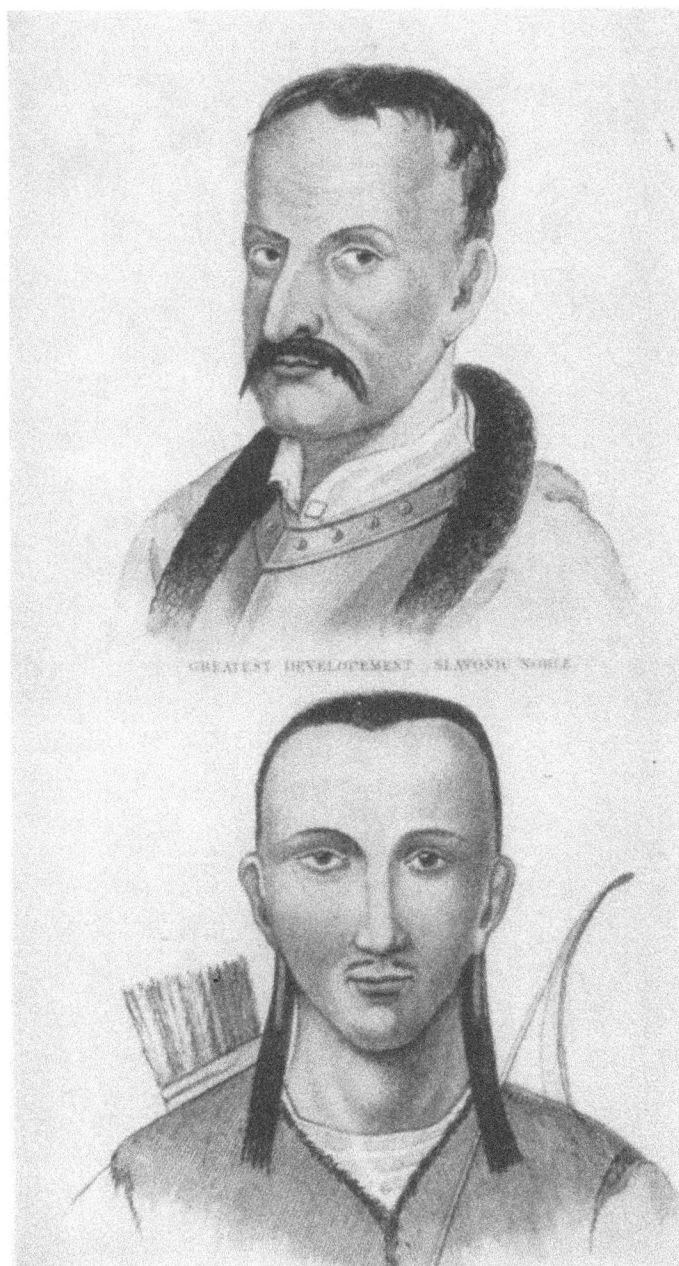

1. Typical Headforms (from Charles Hamilton Smith, The Natural History of the Human Species, 1851)

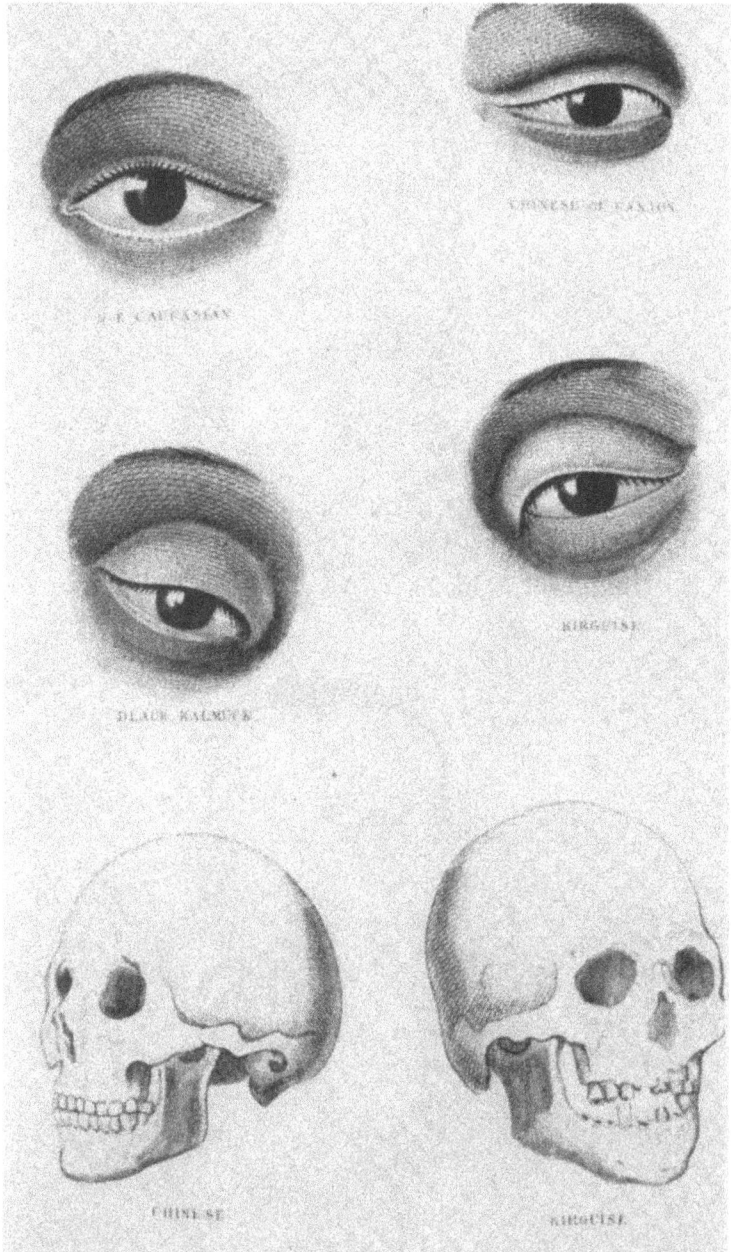

2. Eye Form and Cranial Structure (from Charles Hamilton Smith, The Natural History of the Human Species, 1851)

who produced numerous works on natural history which he illustrated with his own watercolors. With interests in archaeology
and zoology, it was only to be expected that he would eventually
turn to a study of *The Natural History of the Human Species*, which
he published in 1848. Apart from anything else it gave him ample
opportunity to display his artistic talents as he depicted the "typical" human headforms, cranial structure, and eye-form (See Figures 1 and 2).

In this work Smith cautiously kept open the monogenist-
polygenist question; it remained, he said, "a question in systematic zoology, whether mankind is wholly derived from a single
species . . . or produced by the hand of nature at different epochs." But the whole tenor of the work was polygenist as he
specified the "centres of existence of the three typical forms of
man"—"the intertropical region of Africa, for the woolly haired—
the open elevated regions of north-eastern Asia for the beard-
less—and the mountain ranges towards the south and west for
the bearded Caucasian." Indeed, when an American edition of
the work was brought out in 1851 it was under the editorial care
of Samuel Kneeland, a Boston medical naturalist and polygenist
who was known to confirm Morton's measurements of Hindu
crania. In an introduction that amounted to nearly a quarter the
length of the whole book, Kneeland provided a detailed review of
scientific works on race, ever biasing his judgments towards
polygenism, and confirming the orthodoxy of Agassiz-type creationism. What is noticeable here, moreover, is Kneeland's readi-
ness to employ both creationist and naturalistic language to le-
gitimate his convictions about racial inferiority. What races were
by *creation* and by *nature* could not be indefinitely modified, for
degeneration inevitably brought reversion to type. And yet when
monogenists like Rev. Thomas Smyth used scriptural authority to
support that position, Kneeland was only too quick to insist that
this was more the language of "the polemic theologian . . . than
the scientific naturalist, and ethnologist; and, however appropriate in other places, is quite irrelevant on the subject of the origin
of mankind." As for Charles Hamilton Smith himself, his entire
inventory of the human species was charted without recourse
either to biblical exegesis or to the significance of Adam, and this
further shows that caution must be taken not to equate preadamism as a *theological* construct with polygenism as an *anthropological* thesis.[18]

---

[18] Lieut.-Col. Chas. Hamilton Smith, *The Natural History of the Human Species, Its Typical Forms, Priméval Distribution, Filiations, and Migrations* (Edinburgh: Lizars, 1848), 113, 121.

At the same time there were those determined to protest the doctrinal propriety of polygenetic preadamism. During that same year, 1848, for example, William Frederick van Amringe (1791-1873) set out to show that scripture did not deny polygenism. To be sure, he did not explicitly commit himself to such a scheme, but he was at pains to insist that if the conjectures of the most recent physiological anthropologists were to be verified any theological shock waves could easily be absorbed. As we shall see, this was a strategy that became increasingly popular among orthodox Christian apologists. Indeed, van Amringe had the suspicion that Mosaic history fitted rather well with the presumption of multiple human species of different creative origins. The testimony of the Genesis narrative was just too fragmentary, he felt, and all sorts of questions were left unanswered: where Cain got his wife, who peopled his city, who were the strange "sons of God" who intermarried with the "daughters of men." All of these problems evaporated with the coming of the preadamite.[19]

In Britain the self-same anthropological pressures were felt; but here additional archaeological and philological researches added further fuel to preadamism. The English surgeon John Mason Good, a distinguished linguist equally at home with Arabic, Hebrew, Persian, Russian, Sanskrit, and Chinese, drew on his specialist knowledge of these languages and cultures to query the Mosaic narrative and to explicitly return to the conjectures of Peyrère.[20] Similar concerns were also to the fore in *The Genesis of the Earth and Man* published in 1856. The book was edited, introduced, and endorsed by the distinguished British archaeologist and orientalist Reginald Stuart Poole, Head of the Coin Department at the British Museum and later Professor of Archaeology at University College, London. The author was in fact his uncle,

The quotation from Kneeland's introduction appears on p. 60. Biographical details are available in *Dictionary of National Biography*, s.v. Kneeland's Boston version published in 1851 by Gould and Lincoln included the following addition to the title page: "With a Preliminary Abstract of the Views of Blumenbach, Prichard, Bachman, Agassiz, and Other Authors of Repute on the Subject." Later Oscar Peschel argued that the polygenist belief that different human species "were at once sown broadcast by the Creator, in numbers as vast as swarms of bees" was a departure from sound scientific principle. As he went on, "Any explanation of the present by the past is thus abandoned, although it lies deeply rooted in human nature not to rest satisfied with observed facts until they have been reduced under some law of necessity." His objection, to put it another way, was that polygenism departed from vera causa explanations. See Oscar Peschel, *The Races of Man, and their Geographical Distribution* (New York: Appleton, 1906; first published in German, 1874), 33-34.

[19] William Frederick Van Amringe, *An Investigation of the Theories of the Natural History of Man, by Lawrence, Prichard, and Others Founded Upon Human Analogies; and an Outline of a New Natural History of Man Founded Upon History, Anatomy, Physiology, and Human Analogies* (New York, 1848).

[20] John Mason Good, *The Book of Nature* (London, 1826).

Edward William Lane (1801-1876), an Arabic scholar, lexicographer, and traveler. Lane's intention was to integrate biblical religion with the findings of the pioneer American anthropologists Morton, Nott, Gliddon, and Agassiz. In his mind the need for a suitable hermeneutic to cope with these threats to biblical orthodoxy was a real *desideratum* and he hoped that he himself could do for anthropology what Hugh Miller and Edward Hitchcock had already done for geology. But the need of the hour was even greater than that. Recent research on the historical geography of early migration and on comparative philology also had to be taken into account. Indeed, to sustain his defense of the preadamite, Lane was entirely prepared to launch an attack on Bunsen and Müller's monogenetic account of linguistic development. His own scheme, based on the discrimination of monosyllabic, agglutinate, and amalgamate forms, ran counter to Bunsen's classification, first, in "representing the Semitic stock as in no way derived from a primeval language . . . and second, in representing the Egyptian, which he terms 'Khamitism,' or 'Chamitism', as being, from the first, collateral to the Semitic" (see the contrast between Figures 3a and 3b). The implications of this polygenetic taxonomy were, of course, obvious: "there have existed Pre-Adamites of our species."[21]

While Lane's confidence in the preadamite theory was both bolstered and directed by his personal researches on geology, physical anthropology, philology, and early history, his ultimate motivation was to preserve the integrity of Old Testament chronology. Only by viewing Adam as a *recent* newcomer on the stage of human history could the authority of science and scripture remain unimpeached. As he himself put it:

Some men of science and learning, holding the common belief that Adam was the first of our species, have confessed themselves to be compelled, by geological and other considerations, to adopt [the] opinion . . . that he was created twenty thousand years, or more, before the Christian era; and the early Biblical

---

[21] Reginald Stuart Poole (ed.), *The Genesis of the Earth and of Man: or the History of Creation, and the Antiquity and Races of Mankind, Considered on Biblical and Other Grounds* (London, 1856), 200-201. In Poole's opinion Bunsen's schema was conditioned by his prior commitment to monogenism. "Had it not been that Bunsen had started with a fixed persuasion that all languages had one origin" Poole observed, "he would surely not have come to the conclusion he has reached." ibid., xviii. Bunsen's and Müller's arguments, together with their diagrammatic depiction of linguistic evolution from common anti-diluvian roots, are discussed in Adam Kuper, "The Development of Lewis Henry Morgan's Evolutionism," *Journal of the History of the Behavioral Sciences*, 21 (1985): 3-22; Thomas R. Traumann, *Lewis Henry Morgan and the Invention of Kinship* (Berkeley: University of California Press, 1987), 218-220. Müller's views of language were such, however, that he maintained that there was a difference in kind, not merely in degree, between humans and other primates. See Elizabeth Knoll, "The Science of Language and the Evolution of Mind: Max Müller's Quarrel with Darwinism," *Journal of the History of the Behavioral Sciences*, 22 (1986): 3-22.

*(Monosyllabic.)*
PRIMEVAL LANGUAGES.

*(Agglutinate.)*
NIGRITIAN

*(Agglutinate.)*
TURANIAN

*(Amalgamate)*
PROTOTYPE OF THE SEMITIC.

| *(Agglutinate.)* HAMITIC. Cushitic : | *(Amalgamate.)* SEMITIC, OR SYRO-ARABIAN. Hebrew and Phœnecian : Aramaic (*i.e.* Chaldee, Samaritan, Nabathæan, and Syriac) : Arabic, commonly so called : and the two extreme branches of the Semitic stock ; namely, | *(Amalgamate.)* JAPHETIC, OR IRANIAN, OR INDO-EUROPEAN. Celtic : Tracian, or Illyrian : Armenian : Arian : Hellenico-Italic : Slavonic : Lithuanian : and Teutonic. |
|---|---|---|
| Egyptian : and several other Languages, of the Eastern and Northern Regions of Africa. | | |

Himyeritic | and Assyro-
and Ethiopio, | Babylonian.

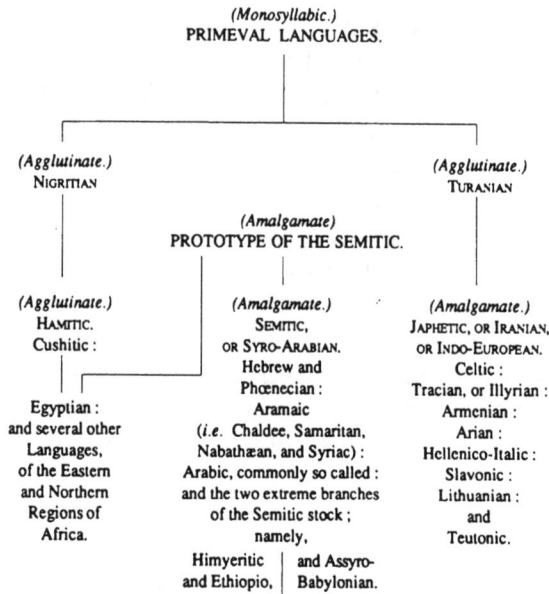

3a. Lane's Dual Origin Model of Philological Development (redrawn from Reginald Stuart Poole (ed.), The Genesis of the Earth and Man, 1860)

history has been enormously distorted to accommodate it to this belief. But our own opinion, that the Bible itself indicates the existence of Pre-Adamites, relieves us from the necessity of requiring a more extended Biblical chronology than that which appears to be advocated by most of the best judges in the present day.

Lane made it abundantly clear that he was sure Adam—"the first individual of a new variety of a species which had universally sinned but not become extinct"—was created by divine fiat, not by any natural development. In his mind, the first human beings, the preadamites, had come into being in the valley of the upper Nile, from which location they diffused through Africa and Asia. The Adamic pair, by contrast, were created to be the ancestors of the Mediterranean peoples. Thereby the views of those "eccentric philosophers" who suggested that "Adam and his wife were Negroes" were scotched.[22]

Although several of the people we have discussed turned to a secular preadamism for racial-scientific reasons, Lane's espousal

---

[22] Poole, *Genesis of Earth and Man* 46, 110, 142-143. Later George Gliddon approvingly and extensively quoted from this work observing that it "augurs well for ethnological progress in Great Britain," and pointing out that it continued the argument earlier put forward by Peyrère. Geo[rge] R. Gliddon, "The Monogenists and the Polygenists: Being an Exposition of the Doctrines of Schools Professing to Sustain Dogmatically the Unity or the Diversity of Human Races; With an Inquiry into the Antiquity of Mankind Upon Earth, Viewed Chronologically, Historically, and Palaeontologically," in J.C. Nott and George R. Gliddon, *Indigenous Races of the Earth* (Philadelphia, 1857), 415.

## LIVING LANGUAGES.

| | | | |
|---|---|---|---|
| **POLITICAL STAGE.** | Concentration of *Chinese.* | **Northern Branch:** Concentration of the *Tungusic.* / Concentration of the *Mongolic.* / Concentration of the *Turkic.* / Concentration of the *Finnic.* (Scattered languages: Bask, Samoïedic, Caucasic.) — **Southern Branch:** Concentration of the *Taïc.* / Concentration of the *Malaïc.* / Concentration of the *Bhotiya* (Gangetic and Lohitic). / Concentration of the *Tamulic.* | **Semitic Nucleus.** National idiom of *Africa, N.W.* / ,, *Egypt.* / ,, *Babylon.* / National idiom of *Arabia.* / ,, *Aram.* / ,, *Palestine.* — **Arian Nucleus.** National idiom of the *Indic branch.* / ,, *Iranic* ,, / ,, *Celtic* ,, / ,, *Italic* ,, / ,, *Hellenic* ,, / ,, *Windic* ,, / ,, *Teutonic* ,, <br> **AMALGAMATION.** |
| **NOMADIC STAGE.** | | **A G G L U T I N A T I O N.** | |
| **FAMILY STAGE.** | **J U X T A P O S I T I O N.** | | |
| **ANTE-DILUVIAN.** | **R O O T S.** | | |

3b. Bunsen's Model of Linguistic Evolution (from C.C.J. Bunsen, Outlines of the Philosophy of Universal History, 1854)

of the theory was much more an apologetic move to keep science and Christianity in conceptual tandem. And this, as we shall now see, tended to be increasingly characteristic of those who warmed to the preadamite theory. Where the idea of a historical Adam was taken seriously, pre-*Adam*ism (in contrast with mere polygenism) provided one possible strategy for maintaining the harmony between the Bible and science.

## Reconciling Ethnology and Theology

Throughout the middle decades of nineteenth century Britain the preadamite theory continued as part of the conventional discourse of the new sciences of anthropology and ethnology. This is plain, for example, from a reading of the proceedings of the Ethnological Society of London. It had originally come into being in the 1840s, at the instigation of Thomas Hodgkin, as an offshoot of the Aborigines Protection Society, and it concerned itself more with the scientific study of the natural history of the human species than with humanitarianism.[23] But it had gone into decline during the 1850s and was later reinvigorated by a group of anthropologists and archaeologists representing the newer trends in their subjects. Soon, however, cracks began to appear, and friction with the older Quaker-humanitarian element in the Ethnological Society resulted in the formation of the breakaway Anthropological Society in 1863.[24] The details of this institutional chapter in the history of British Anthropology need not detain us here, save to say that the new society represented a tougher polygenetic and racialist line of thinking. Thus, as we shall see, the writings of certain key members, like James Hunt and Robert Knox, display a more secular polygenism that should, perhaps, not be conflated with the more religiously inspired notion of preadamism. In the present context, however, I only intend to show that the preadamite theory provided a convenient vocabulary for scientific discourse about the origins of the human species in the early Victorian period.

From the first issue of the second series of the society's *Transactions*, the name of the aging John Crawfurd (1783-1868)—a close associate of Roderick Murchison—featured regularly. Crawfurd, who played a major role as president and vice-president of the Ethnological Society, was an Indian Army doctor turned Orientalist and an East India Company resident at Singapore. He steadfastly promoted polygenism from a variety of perspectives ranging from physical anthropology to philology. "That the many separate and distinct races of man . . . are originally created spe-

---

[23] On the origins of the Ethnological Society, see Stocking, *Victorian Anthropology*, 240-245.

[24] The story of the formation of the Anthropological Society of London has frequently been told. See, John W. Burrow, "Evolution and Anthropology in the 1860s: The Anthropological Society of London 1863-1871," *Victorian Studies*, 7 (1963): 137-154; George W. Stocking Jr., "What's In a Name? The Origins of the Royal Anthropological Institute, 1837-1871," *Man. The Journal of the Royal Anthropological Institute*, 6 (1971): 369-390; Ronald Rainger, "Race, Politics and Science: The Anthropological Society of London in the 1860s," *Victorian Studies*, 22 (1978): 51-70; Stocking, Victorian Anthropology, 247-254.

cies, and not mere varieties of a single family," he affirmed, ". . . there are many facts to show."[25] Understandably he accorded a minimal role to the influence of physical and social environment on human physiology: neither form, nor stature, nor complexion were subject to climatic modification or to the effects of diet.[26] By this assertion Crawfurd intended to make it clear that race was the product of creation, not of climate, still less of evolution.[27] To be sure, it had "pleased the Creator—for reasons inscrutable to us—to plant certain fair races in the temperate regions of Europe, and there only, and certain black ones in the tropical and sub-tropical regions of Africa and Asia, to the exclusion of white ones, but it is certain that climate has nothing to do with the matter." As he further explained:

When man was first called into existence, we may conjecture that he was planted over the earth in many small families or groups, often consisting of distinct species, or when not so of a single species, adapted to a variety of climates. . . . In such an isolated state, each little group would construct its own separate language . . . and hence the multiplicity of tongues, and the variety of words and structures.

So committed was Crawfurd to this scheme that he dismissed Prichard's monogenism as a "monstrous supposition" hardly worthy of "serious refutation."[28] Still, for all his comments on racial inferiority and superiority, Crawfurd steadfastly set his face against slavery and in so doing revealed that polygenism was not invariably cast in the role of providing scientific legitimation for the ideology of slavocracy.

Crawfurd was certainly not alone among members of the Ethnological Society in his belief in the creation of non-Adamic humans. On the basis of his findings in Egyptian ethnology, for

---

[25] John Crawfurd, "On the Classification of the Races of Man," *Transactions of the Ethnological Society of London*, new series, 1 (1861): 354-378, on pp. 355-356. Crawfurd's interactions with Murchison feature in Robert A. Stafford, *Scientist of Empire. Sir Roderick Murchison, Scientific Exploration and Victorian Imperialism* (Cambridge: Cambridge University Press, 1989), 45, 53, 134f., 148f.

[26] John Crawfurd, "On the Effects of Commixture, Locality, Climate, and Food on the Races of Man," *Trans. Ethnol. Soc. Lond.*, new series, 1 (1861): 76-92; see also idem, "On the Connexion between Ethnology and Physical Geography," *Trans. Ethnol. Soc. Lond.*, new series, 2 (1863): 4-23.

[27] Crawfurd's virulently anti-Darwinian sentiments with respect to ethnology are apparent in "On the Theory of the Origin of Species by Natural Selection in the Struggle for Life," *Trans. Ethnol. Soc. Lond.*, new series, 7 (1869): 27-38.

[28] "Classification," *Trans. Ethnol. Soc. Lond.*, new series, 1 (1861): 365, 371, 361. It should be pointed out that Crawfurd felt that language, because it was acquired rather than innate, was itself not a safe guide to race. In a communication "On the Aryan or Indo-Germanic Theory," *Trans. Ethnol. Soc. Lond.*, 1 (1861): 268-286, he accordingly elaborated criticisms of Müller's monogenetic thesis that traced all languages to Sanskrit sources. See also his paper "On Language as a Test of the Races of Man," *Trans. Ethnol. Soc. Lond.*, new series, 3 (1865): 1-8.

example, Reginald Stuart Poole explicitly endorsed the double-Adamism of his uncle in a paper read before the Society in June 1862. But he hastened to dissociate himself from the nasty racialist implications that some had drawn from polygenism. "No ethnological theory or fact can shake the first principles of Christian morality," he insisted, wryly adding that belief in the common ancestry of the human species had not prevented good monogenists in America's southern states from abusing blacks. In fact he went on to insist that "the very knowledge that an inferior race had a separate origin, should rather arouse our generosity than awaken our dislike, and be an incentive to liberality rather than to tyranny."[29] The following year the Reverend Frederic W. Farrar presented the Society with superabundant scientific testimony to the fixity of type and the impotence of climate, urging his readers to draw from his catalogue "such inferences as appear . . . to be most truthful and logical."[30]

The inferences to be drawn were hardly uniform, and different Society members reacted in different ways. On the one hand, there were those who continued to believe that even if polygenism were to be verified, it would leave the question of the specific unity of humankind entirely untouched; on the other, there were those who pressed polygenism into the service of the harshest of racial ideologies. Consider first the case of Robert Dunn, a clinical psychologist and Fellow of the Royal College of Surgeons who repeatedly argued the case for the unity of mankind from anthropometric measurements of human crania, from the tegumentary differences between races, as well as from psychology. As might be expected, Dunn considered environment to be a large influence in modifying skin color and hair character, and went on to adduce unitary arguments from hybrid fertility. He did make it clear nonetheless that while "the unity of the species by physiological and psychological evidence has been established, the *"quaestio vexata"* still remains: Have there been more creations than one of the same genus, more Adams and Eves than one single pair?" He himself said that he sided with the monogenism of Edward Forbes, R.G. Latham, William B. Carpenter and Prichard against Louis Agassiz; the supposition of the creation of several distinct *"protoplasts,"* one for each region of the globe, was not required to account for the geographical distribution of race. Indeed,

---

[29] Reginald Stuart Poole, "The Ethnology of Egypt," *Trans. Ethnol. Soc. Lond.* new series, 2 (1863): 260-264, on p. 264.

[30] Frederic W. Farrar, "Fixity of Type," *Trans. Ethnol. Soc. Lond.* new series, 3 (1865): 394-399, on p. 399.

he repeated this same contention a few years later when he told the Society that the "universal brotherhood" of man stemmed from God's making "of one blood all the nations of the earth," not necessarily from lineal descent from an original single pair.[31]

Others, however, were rather less prepared to leave the monogenist-polygenist question so open-ended. Chief among these were James Hunt (1833-69), whose departure to found the polygenist-oriented Anthropological Society of London is well known; and Robert Knox (1791-1862), the Edinburgh anatomist, whose racialist philosophy had profoundly influenced Hunt. Their rather more secular polygenetic version of racial origins was, certainly, not preadamite in any theologically significant sense. And herein yet again lies the danger of too narrowly equating preadamism or coadamism with polygenism. Hunt, for example, was no happier with those committed to Moses than with those enamored of Darwin. There was, to him, little difference between "a disciple of Darwin and a disciple of Moses—one calls in natural selection with unlimited power, the other calls in a Deity provided in the same manner."[32] Still, the conceptual allegiances of Hunt and Knox continued to illustrate the kinds of historical and scientific explanation to which advocates of Preadamism were of necessity drawn. Hunt, for example, had already steadfastly set his face against acclimatization of any kind when he spoke to the Ethnological Society in February 1862 on the subject of "Ethno-Climatology; or the Acclimatization of Man." Here he drew on the demographic statistics of mortality compiled by medical practitioners such as Ranald Martin and A.S. Thomson, to resist the notion of human cosmopolitanism.[33] Hunt for one, acknowledged the importance of the study of acclimatization to the colonial interests of Great Britain, but his concern stemmed from the degeneracy that he felt accompanied the transplantation of races and individuals to different climatic regimes.[34]

---

[31] Robert Dunn, "On the Physiological and Psychological Evidence in Support of the Unity of the Human Species," *Trans. Ethnol. Soc. Lond.*, 1 (1861): 186-202, on p. 201; idem, "Some Observations on the Psychological Differences Which Exist among the Typical Races of Man," *Trans. Ethnol. Soc. Lond.* 3 (1865): 9-25, on p. 11. Dunn read this paper to the Society in 1863, having already delivered it at the 1862 Cambridge meeting of the British Association. See also idem, "Some Observations on the Tegumentary Differences Which Exist among the Races of Man," *Trans. Ethnol. Soc. Lond.* 1 (1861): 59-71. In thus citing Forbes, Dunn may have given the impression that Forbes believed in a single center of creation. This is not so. See Ernst Mayr, *The Growth of Biological Thought. Diversity, Evolution, and Inheritance* (Cambridge, Mass.: Harvard University Press, 1982), 444.

[32] Quoted in Stocking, *Victorian Anthropology*, 249.

[33] These individuals are discussed in Livingstone, "Human Acclimatization" (cit. n. 1).

[34] James Hunt, "On Ethno-Climatology; or the Acclimatization of Man," *Trans. Ethnol. Soc. Lond.*, new series, 2 (1863): 50-79.

Hunt's diagnosis was informed throughout by the writings of Robert Knox who had published in 1850 *The Races of Men*. Knox, a British anatomist and former army surgeon, had turned to the study of ethnology in the 1840s in the aftermath of the infamous Burke and Hare scandal involving the procuring of cadavers for class dissection. Here Knox provided a guide to a variety of races—Jewish, Coptic, Germanic, Celtic, Slav, Black, and so on, wedding his diagnosis to a mélange of rigid anti-acclimatization-ism, hereditarianism and polygenism, drawing all the while on the continental tradition of transcendental biology. Archetypal plans, not external environment, best explained organic diversity. So firmly committed was Knox to the idea that racial groups were structurally fitted for specific regional environments that he went so far as to claim that no race could permanently change its ter-restrial zone without degeneracy. For his efforts Knox was re-warded with election to membership in the *Société d'Anthropologie de Paris* in 1861 and with Ralph Waldo Emerson's grudging ac-ceptance of his "unpalatable conclusions" as "charged with pun-gent and unforgettable truths."[35]

The emergence of a more secular polygenism—and witnessed in the materialistic radicalism of Knox—as the second half of the nineteenth century dawned, certainly did not mean the demise of the theological preadamite from the scientific scene. At the An-thropological Society itself much was made of Peyrère and his role in the intellectual prehistory of polygenism. Thomas Bendysche, for instance, one of the Society's vice-presidents, outlined Peyr-ère's crucial contributions in his memoir on the "History of An-thropology." A subsequent contributor, writing as "Philalethes," set out to rescue Peyrère no less from oblivion than from igno-miny, and compared Peyrère's perfidious treatment at the hands of the self-designated orthodox to the virulence Samuel George

---

[35] Robert Knox, *The Races of Men: A Philosophical Enquiry into the Influence of Race over the Destinies of Nations* (London: Renshaw, 2nd ed., 1862), 106-145. Biographical details are available in Henry Lonsdale, *A Sketch of the Life and Writings of Robert Knox, the Anatomist* (London: Macmillan, 1870). See also the discussion in Livingstone, "Human Acclimatiza-tion" (cit. n. 1), 367-368. Knox contributed several papers to the Ethnological Society of London on subjects ranging from Assyrian marbles and their place in history and art, to observations on human crania. Ralph Waldo Emerson's words come from *The Conduct of Life*, volume 6 of Emerson's *Complete Works* (London: George Routledge and Sons, n.d.), 21. Various aspects of Knox's thinking are discussed in Philip F. Rehbock, *The Philosophical Naturalists: Themes in Early Nineteenth-Century British Biology* (Madison: University of Wis-consin Press, 1983), chapter 2 on "Robert Knox: Idealism Imported"; and Evelleen Rich-ards, "The 'Moral Anatomy' of Robert Knox: The Interplay between Biological and Social Thought in Victorian Scientific Naturalism," *Journal of the History of Biology*, 22 (1989): 373-436.

Morton had suffered from critics like John Bachman and Thomas Smyth.[36]

At the same time there appeared during the 1850s and 1860s a number of works defending various preadamite schemes from a range of different perspectives. Take as a single instance that of a certain Isabella Duncan. Her *Pre-Adamite Man*, which appeared in several editions, was a work explicitly modeled on Peyrèrean lines and presented as a means of making sense of the two distinct accounts of human creation recorded in the early Genesis narratives. Much of the work was a scientific exposition of the creation story, drawing on the geological works and harmonizing schemes of men like Thomas Chalmers, John Pye Smith, and Hugh Miller, and to that extent it was a standard contribution to scriptural geology. But Isabella Duncan also had extra-scientific, not to say fanciful, concerns. She affirmed the existence of a human race before Adam, with a history that had run its course prior to the appearance of the Adamic family. But she then went on to insist that there were no archaeological traces of this preadamic civilization for the simple reason that the preadamites had all the attributes of angels! This particular rendition of the preadamite theory involved the author not only in some tortuous exegesis of various Old Testament passages, but also in the claim that the major geological catastrophes of the past were to be correlated with the fall of the preadamite angels from their original state of perfection.[37]

Conventional scientific concerns were much more to the fore in *Adam and the Adamite*, an 1864 volume tellingly subtitled "the Harmony of Scripture and Ethnology" and written by Dominick M'Causland Q.C. (1806-1873). During his time on the legal circuit of north-west Ireland, for which he was eventually appointed

---

[36] Thomas Bendysche, "The History of Anthropology," *Memoirs Read before the Anthropological Society of London*, 1 (1863-1864): 335-420; Philalethes, "Peyrerius and Theological Criticism," *Anthropological Review*, 2 (1864): 109-116.

[37] [Isabella Duncan], *Pre-Adamite Man or the Story of our Planet and Its Inhabitants told by Scripture and Science* (London: Saunders, Otley, & Co., 2nd ed., 1860). This work was remarkably well received in the press at the time, even if reviewers could not go the whole way with the theory, as is evident from the review extracts reprinted in the first pages of this second edition. Among other advocates of preadamism at the time were Griffin Lee, *Pre-Adamite Man: The Story of the Human Race from 35,000 to 100,000 Years Ago!* (New York: Sinclair Tousey, 1863); and Nemo [W. Moore], *Man: Palaelithic, Neolithic and Several Other Races, not Inconsistent with Scripture* (Dublin: Hodges, Foster, 1876). John Harris's, *The Pre-Adamite Earth: Contributions to Theological Science* (Boston, 1849), was appropriately named for it dealt with the preadamite earth and not with preadamite humans. As such it was another contribution to scriptural geology. On the latter subject see Milton Millhauser, "The Scriptural Geologists: An Episode in the History of Opinion," *Osiris*, 11 (1954): 65-86.

crown prosecutor under the administration of Lord Derby in 1858-59, M'Causland had found time to compose a number of religious works. In the volume under consideration, his aim was to uphold the detailed accuracy of scripture in the face both of scientific challenges and of the critical spirit of Bishop Colenso and *Essays and Reviews*. From the outset M'Causland freely admitted that the biblical text "was not written for our instruction in any of the physical sciences." But this did not for a moment mean that science and scripture dealt with different spheres and remained cognitively dissonant; it was, he insisted, a "mischievous error to seek to divide Scripture and science and separate the deep mysteries and important truths of revealed religion from the rich treasures of philosophy which God has provided for our instruction." Plainly, the sacred oracles "ought not to contradict any fact that is authenticated by scientific research," thus, to preserve the accuracy of Mosaic primeval history, M'Causland turned to the preadamite. The findings of the geologists, historians, archaeologists, philologists and ethnologists all came within this purview and pressed him to concede that if Adam was to be taken as the progenitor of all mankind, the chronology of the Old Testament would have to be abandoned, "and all that is written in the Book of Genesis of antediluvian members of his family must be treated as the fanciful speculations of some visionary mythologist." Happily such a judgment would be too precipitate, for the biblical Adam was only the last of a series of human types that God had created. Superior to his forebears, Adam's appearance *ex nihilo* in the image of God was recorded for all posterity, although in a short time the pure Adamic line would be sullied through intermarriage with preadamite stock. In M'Causland's case, as earlier with Lane, the preadamite theory was deployed to safeguard the integrity of scripture from the assaults of higher critics—a quite remarkable reversal of its earlier engagement as a source of skepticism.[38]

That the polygenist thesis was thus finding favor with Christian apologists and scientific racists alike certainly does not mean that monogenist adherents to the traditional Adamic narrative had disappeared. Throughout the middle decades of the century the conventional monogenist history continued to be defended

---

[38] Dominick M'Causland, *Adam and the Adamite; or, the Harmony of Scripture and Ethnology* (London: Richard Bentley, 2nd edition, 1868), 153, 3, 153, 156. M'Causland's distaste of *Essays and Reviews* was no doubt particularly stimulated by C. W. Goodwin's contribution "On the Mosaic Cosmology" in which he discarded all concordist schemes—Buckland's, Chalmers's and Miller's. *Essays and Reviews* (London: Longman, Green, Longman & Bros, 1861), 207-253.

by successors of Stanhope Smith and James Cowles Prichard. In America's southern states, for example, some ministers stood out against what they saw as the malign implications of Mortonite polygenism. John Bachman and Thomas Smyth thus defended the unity of the human race on scientific and scriptural grounds by arguing that polygenism was born in infidelity and nurtured in skepticism. Yet this certainly did not imply that they were committed to egalitarianism. The idea of black inferiority was too ingrained for that. Bachman, for example, staunchly defended Southern slavery and argued, on the basis of the biblical curse on Ham, that the black race was designed, and destined, for servitude. For Bachman, polygenism threatened not only the authenticity of scripture, but also the ideological fabric of what he considered Christian civilization. In both cases, elements of the Scottish philosophy also had a key role to play; Bachman argued against several human creations because God did not multiply miracles indiscriminately, while Smyth applied the *vera causa* argument against polygenism.[39] Soon monogenists could also call on the testimony of the Frenchman Armand de Quatrefages, who exerted a formative influence on Isidore Geoffroy Saint-Hilaire and the French tradition of acclimatization.[40]

Quatrefages, even while president of the Société d'Anthropologie, stood out against the majority of French anthropologists who favored Broca's polygenetic views of race and who accounted for organic modification purely by interbreeding. In fact Quatrefages traced the history of polygenism back to its roots in Peyrère's theology noting that "polygenism generally regarded as the result of Free Thought was biblical and dogmatic in origin." The history of monogenism was no less ideologically littered, and thus to avoid the taint of sectarian bias he hastily added that "monogenists are guilty of seeking in religious doctrines arguments in favor of their theory, and anathematizing their adver-

---

[39] John Bachman, *The Doctrine of the Unity of the Human Race Examined on the Principles of Science* (Charleston: Canning, 1850); Thomas Smyth, *The Unity of the Human Races Proved to be the Doctrine of Scripture, Reason and Science with a Review of the Present Position and Theory of Professor Agassiz* (New York: Putnam, 1850). See the discussion in Stanton, *Leopard's Spots*, 123-136. There were, of course, those who used scriptural arguments in favor of slavery. See Thornton Stringfellow, *Scriptural and Statistical Views in Favor of Slavery* (4th edition with additions. Richmond: J.W. Randolph, 1856); George Junkin, *The Integrity of our National Union, vs. Abolitionism: An Argument from the Bible, in proof of the position that believing masters ought to be honored and obeyed by their own servants, and tolerated in, not excommunicated from, the church of God: being part of a speech delivered before the synod of Cincinnati, on the subject of slavery, September 19th and 20th, 1843* (Cincinnati: Printed by R.P. Donogh, 1843); George Dodd Armstrong, *The Christian Doctrine of Slavery* (New York: C. Scribner; Richmond Va.: P.B. Price, 1857).

[40] See Osborne, *Société Zoologique d'Acclimatation*, 283-300.

saries in the name of dogma." As for himself, Quatrefages took
seriously the findings of natural history and used the writings of
figures like Buffon, Lamarck, and Cuvier to support his critique
of Agassiz's centers of creation. Moreover he acknowledged that
stances taking up such scientific questions as variation, migration, and acclimatization were invariably conditioned by commitments to monogenism or polygenism.[41]

It was the continued survival of scientific monogenism among
individuals like these that helped many keep preadamism firmly
in forbidden theological territory. Thus the traditional adamic history continued to be represented in both popular and more serious theological works. In a lecture delivered in Exeter Hall in
December 1848, for example, Rev. William Brock directed
Y.M.C.A. minds to the subject of "The Common Origin of the
Human Race." Patching together a pastiche of popular philology,
physical anthropology, and psychology, the rhetorician urged the
common origin of mankind upon his young hearers. With exuberant enthusiasm he concluded his homily with an appeal to
human solidarity as the basis for evangelistic endeavor.[42] It was a
similar story for Elisha Noyce, too, whose Outlines of Creation for
juvenile readers likewise held up monogenism as the only respectable scientific option.[43]

In Scotland, evangelical Calvinists were certainly no more inclined toward polygenism. In 1850, for example, Hugh Miller,
though widely applauded by preadamite partisans for his harmonizing of Genesis and geology, made it clear that he would have
none of the fashionable polygenetic talk. To him the whole soteriological structure of Christianity rested on the assumption of the
specific unity of the human race and any notion that "Adams and
Eves [were] many" had to be ruled out, simply because, as he put
it, the "second Adam died but for the descendants of the first."
Miller's essay in fact was an extended commentary on Thomas
Smyth's volume on The Unity of the Human Races, and while he
took the same anthropological line, he felt that Smyth had perhaps been too precipitate in arguing that both blacks and blue-
eyed Goths could not have descended naturally from a common
origin. To be sure, supernatural intervention could not be ruled

[41] A. de Quatrefages, The Human Species (London: Kegan Paul, 2nd ed., 1879), 31, 33. See also his "Histoire Naturelle de l'Homme. Unité de l'Espèce Humaine," Revue de Deux Mondes, 30 (1860): 807-833.

[42] William Brock, "The Common Origin of the Human Species," in Lectures to Young Men; Delivered before the Young Men's Christian Association, in Exeter Hall, From November 21, 1848, to February 6, 1849 (London: Jones, 1849), 115-143.

[43] Elisha Noyce, Outlines of Creation (London: Ward & Lock, 1858), 327-338.

out *a priori*; but with a typically Calvinist suspicion of multiplying miracles gratuitously, he proposed that the horns of the dilemma would be removed if it were assumed that Adam was neither black nor white, but a "mingled Negroid and Caucasian type." If so, then "neither the Goth nor the negro would be so extreme a variety of the species as to be beyond the power of natural causes to produce."[44]

Miller's antipathy to the plural origins of mankind was shared by his Free Church of Scotland colleague, John Duns (1820-1909), Professor of Natural Science at New College, Edinburgh from 1864 until 1903, as successor to the distinguished John Fleming. The relationship between science and faith was Duns's scholarly passion and it was only to be expected that the challenge of the newest anthropology would sooner or later come to the surface in his writings. In *Science and Christian Thought and Biblical Natural Science*, both of which appeared in the mid 1860s, he thus tackled the question of the antiquity of the human species. Throughout, Duns, in the tradition of Calvinistic epistemology, made it clear that his science was grounded in Christian presuppositions and that the claims of the biblical world view should be taken on the principle of initial credulity, at least until "its insecurity is clearly and triumphantly shown." The scriptural record, taken *prima facie*, plainly indicated a monogenist position, and Duns only found this confirmed as he perused the arguments and sifted the data of natural historians, philologists, and anthropologists alike. Moreover, Duns found himself revolted no less by the evolutionary musings of Carl Vogt than by the ethical relativism of continental materialists and by the racist undercurrents among the American Mortonites, and he thus set himself the task of refuting the underlying naturalistic philosophy.

Duns's survey of pertinent scientific literature was prolix; but his familiarity with the sources of the newest science did nothing to weaken his resolute adherence to the Adamic world-picture. Again following the well-worn path of Prichard and Latham, he looked to the "power of *habitat*, climate and the like" to produce as great variations among the peoples of the world as among the

---

[44] Hugh Miller, "Unity of the Human Races," (1850) in *Essays. Historical and Biographical, Political and Social, Literary and Scientific* (Edinburgh: William P. Nimmo, 4th ed., 1870), 387-397, on pp. 392, 396. Later Miller seems to have rejected this view when he asserted: "I do not see how we are to avoid the conclusion that this Caucasian type was the type of Adamic man." Hugh Miller, *Testimony of the Rocks; or, Geology in Its Bearing on the Two Theologies, Natural and Revealed* (Edinburgh: Thomas Constable, 1857), 251-252.

lower animals, thereby paradoxically connecting up animal and human in a single explanatory network. "Is there greater unlikeness between the head of the negro, the aboriginal Australian, the European, and the Hindoo," he rhetorically asked, "than there is between the head of the grey-hound and that of the mastiff or the bulldog?" Science, it seemed, was on the side of scripture—as was morality, for Duns slyly remarked that it was only to be expected of the Morton-Nott-Gliddon brigade that in their portrait of universal race history they "would assign a foremost place to the families to which they themselves belong." Interwoven with the fabric of Duns's diagnosis were theological threads. Thus while scrutinizing archaeological and geological evidence of human antiquity (from Danish kjökken-nöddings [or kitchen middens], lake dwellings, peat deposits from Scotland and swamp cypress in Mississippi, not to mention other paleobotanical data) and consistently reinterpreting it to shorten the time scale, he spelled out in detail the biblical objections to the preadamism of the recently published *Genesis of the Earth and Man* and *The Pre-Adamite Men*, as he focused particularly on the Pauline theology of original sin.[45]

Even while monogenists were continuing to question the preadamites' theological and scientific legitimacy, there were those who could see their functional value as a last line of religious defense against skeptical science. Certainly such observers, as

[45] John Duns, *Science and Christian Thought* (London: Religious Tract Society, [circa 1866]), 207, 301, 292; idem, *Biblical Natural Science, Being the Explanation of All the References in Holy Scripture to Geology, Botany, Zoology, and Physical Geography* (Edinburgh: William Mackenzie, [1863-66]), 534-538. A few scattered biographical fragments are available in Hugh Watt, *New College Edinburgh. A Centenary History* (Edinburgh: Oliver & Boyd, 1946), 55-57, 248-249. Duns's final reference was probably to the argumentative work by Griffin Lee of Texas, *Pre-Adamite Man. The Story of the Human Race from 35,000 to 100,000 Years Ago!* (New York: Sinclair Tousey, 1863) which set out the standard evidence from linguistics, anthropology, archaeology, paleontology, and ancient history, for the existence of humans long before the biblical chronology for Adam would allow. In contrast to Duns, Lee opted for a plurality of human species and rejected climatic accounts of racial differentiation. Accordingly, commenting on the black race, he noted: "Climate, therefore, has nothing to do in making him the hue he is; for in Africa, beneath the torrid sun, and in cold Icelandic regions, he and his children retain the same complexion. . . .He must, therefore, be a distinct species of the genus homo; for, differing so totally from the sons of Adam, he could not have descended from the same source, and is consequently, and beyond all doubt, a Pre-Adamite Man." ibid., 67-68. Yet Lee did insist that, though a different species, the black race was destined "for power and greatness," and dedicated the volume to Abraham Lincoln. Lee's book was scathingly reviewed, indeed lampooned in a lengthy review by Rev. R. Weiser for the *Evangelical Quarterly Review*, 17 (1866): 222-236. For some reason that I have not been able to ascertain, Lee's book also appeared in the same year of publication under the name of Dr. Paschal Beverley Randolph, and with the slightly modified title of *Pre-Adamite Man: Demonstrating the Existence of the Human Race upon this Earth 100,000 Thousand Years Ago!* (Toledo, Ohio: Randolph Publishing Company, 1863). Perhaps it was because, according to the Preface to the Fourth (1888) Edition, "the book has had mountains almost insurmountable in its path."

often as not, remained uncommitted to preadamism; but they wanted to leave the scheme as a viable strategy should science's darker speculations come to be verified. Now in the 1870s, especially in the wake of M'Causland's treatise, some theological conservatives toyed with making their peace with the preadamite. To be sure, such anticipatory tactics were not altogether new. John Pye Smith (1774-1851), Congregationalist clergyman and geologist, had mapped out as early as 1840 the theological path to be followed should polygenism come to be vindicated. *If* Morton's cause was to be sustained, he explained, preadamism remained an acceptable option, for after all such a doctrine denied neither God's special creation of Adam, nor his theological significance as "a figure of Him that was to come." Besides, he told his readers that Isaac de la Peyrère and Edward King had already shown how "some difficulties in the scripture-history would be taken away; such as—the sons of Adam obtaining wives, not their own sisters;—Cain's acquiring instruments of husbandry, which have been furnished by miracle immediately from God upon the usual supposition."[46]

Now, some thirty years later on, another Congregationalist clergyman, Joseph P. Thompson, an editor of the *Independent* and co-founder of the *New Englander*, made it clear in *Man in Genesis and Geology* that although the preadamite theory of Peyrère himself was "open to some serious objections, it serves to show one possible way in which the Bible and Science may yet be harmonized upon the question of the antiquity of man and the unity of the race. It may prove eventually that there is in this brief record in Genesis a margin [sic] for all the discoveries of Science."[47] The following year Enoch L. Faucher, writing in *Scribner's Magazine*, posed the question "Was Adam the First Man?" Like M'Causland, he was anxious to rehabilitate biblical chronology in the light of recent archaeological investigations. Two schemes of reconciliation, he reported, were currently available, namely, M'Causland's and Lane's Preadamism, and the Duke of Argyll's *Primeval Man* which pressed Adam's antiquity much farther back in time.

---

[46] John Pye Smith, *On the Relation between the Holy Scriptures and Some Parts of Geological Science* (London: Jackson and Walford, second edition, 1840), 391-392.

[47] Joseph P. Thompson, *Man in Genesis and Geology; or, the Biblical Account of Man's Creation, Tested by Scientific Theories of his Origin and Antiquity* (New York: S.R. Wells, 1870), 106-107. It was perhaps appropriate that Thompson should dedicate this volume to James Dana in recognition of his attainments in "illustrating the harmony of truth in the works and word of God." Thompson was also the author of *The Question of Races in the United States* (Glasgow: Anderson, 1874). In the interim Agassiz's theories and the arguments of Peyrère were defended by the Unitarian minister, N.L. Frothingham in "Men before Adam," *The Christian Examiner and Religious Miscellany*, 50 (1851): 79-96.

As Faucher perused the anthropometric, philological, archaeological and geological literature, not to mention the biblical documents themselves and the writings of commentators like Thompson, he was forced to conclude that the preadamite theory could indeed help reconcile the declarations of Scripture with the revelations of science and should probably be accepted in the absence of any coherent alternative.[48]

In that same year, 1871, a volume entitled *The Beginning: Its When and Its How* also made an appearance. Written by Mungo Ponton, a Fellow of the Royal Society of Edinburgh and a photographic inventor, it was appropriately enough spiced with a series of luscious paleontological illustrations. In his exposition of what he called the sixth creative epoch, Ponton called attention to M'Causland's work and affirmed that "Should the evidence then ultimately prove sufficient to establish the existence of races of men older than the Caucasian, it would not be difficult to reconcile the Hebrew narrative with such a state of affairs"; there was always the preadamite safety-valve. "There may have been a Negro Adam, a Mongolian Adam, and perhaps two or three more besides the Caucasian Adam," he affirmed, and they may have been formed either at the same time, or in a series of separate creations.[49]

Thus far Christian apologists who could see theological potential in preadamism invariably cast the theory in polygenetic terms resorting to the conception of multiple creations. One indeed might suspect that preadamism was just a Christianized version of polygenism. This, however, was not necessarily so. The preadamite theory could be, and was, tailored to meet monogenetic requirements. By so doing its advocates felt it fitted even more comfortably with the Genesis picture. And nowhere was this aspiration more clearly displayed than in the writings of the Michigan geologist, Alexander Winchell.

### Alexander Winchell and Monogenetic Preadamism

Even while the polygenism of the American school of ethnologists was rapidly spreading and Christian apologists were flirting with the preadamite in their continued endeavor to keep alive the marriage of science and religion, Alexander Winchell's magisterial *Adamites and Pre-Adamites* (1878) appeared and this work

---

[48] Enoch L. Faucher, "Was Adam the First Man?" *Scribner's Magazine* 1 (1871): 578-589.

[49] Mungo Ponton, *The Beginning: Its When and Its How* (London: Longmans, Green, 1871), 504-505. Ponton contributed several papers on optical subjects to the *Edinburgh New Philosophical Journal*. See "Ponton, Mungo," *Dictionary of National Biography*, s.v.

that departed materially from both these strands of thought. On
the one hand, Winchell's version of preadamism was thoroughly
*monogenetic* in contrast to the conventional rendering; and on the
other, not content simply to introduce preadamism as a mere
theoretical possibility, he actively embraced the theory as the only
means of maintaining the credibility of biblical chronology.
Winchell (1824-1891) enjoyed a distinguished career in American
science both as a geologist and as a reconciler of science and
religion, and throughout his life retained a typically Arminian
enthusiasm for natural theology.[50] Besides publishing numerous
technical works on petrology and mineralogy, he contributed regularly to the *Methodist Quarterly Review,* and thereby assumed the
role of a purveyor of science to the Methodist fraternity. His re-
hearsal of the teleological argument, however, departed from the
Paleyan norm, for he resorted to the idealistic, or better, homo-
logical version associated with such figures as Richard Owen and
James McCosh[51]—an account that facilitated his adoption of theistic evolution in its neo-Lamarckian guise.[52] By 1874 he was on
record as saying that "the doctrine of evolution" was "clearly the
law of universal intelligence under which complex results are
brought into existence," and within three years he had declared
his support for the neo-Lamarckian mechanism of Edward
Drinker Cope.[53]

---

[50] Winchell's academic career was largely spent at the University of Michigan, where he
variously served as professor of physics and engineering, and later of geology, zoology,
and botany. For a short time he was first chancellor of the University of Syracuse, having
been appointed to that position in 1873, and then in 1876 he moved to a chair at Vanderbilt.
This position was summarily abolished in 1878 for reasons that will be discussed shortly,
and he returned to Michigan in 1879 to serve as Professor of Geology and Palaeontology.
Biographical details are available in F. Garvin Davenport, "Alexander Winchell: Michigan
Scientist and Educator," *Michigan History,* 35 (1851): 185-201; idem, "Scientific Interests in
Kentucky and Tennessee, 1870-1890," *Journal of Southern History,* 14 (1948): 500-521; Paul K.
Conkin, *Gone with the Ivy. A Biography of Vanderbilt University* (Knoxville: University of
Tennessee Press, 1985), 50-51, 60-63. Aspects of his thought are also discussed in John S.
Haller, Jr., *Outcasts from Evolution. Scientific Attitudes of Racial Inferiority, 1859-1900* (Urbana,
Chicago and London: University of Illinois Press 1971); David N. Livingstone, "The His-
tory of Science and the History of Geography: Interactions and Implications," *History of
Science,* 22 (1984): 271-302; and in idem, *Darwin's Forgotten Defenders,* 85-92.
[51] See Peter J. Bowler, "Darwinism and the Argument from Design: Suggestions for a
Re-evaluation," *Journal of the History of Ideas,* 10 (1977): 29-43; and David N. Livingstone,
"The Idea of Design: The Vicissitudes of a Key Concept in the Princeton Response to
Darwin," *Scottish Journal of Theology,* 37 (1984): 329-357.
[52] Winchell's style of natural theology is revealed in Alexander Winchell, *Theologico-
Geology, or, The Teaching of Scripture, Illustrated by the Conformation of the Earth's Crust* (Ann
Arbor: Davis & Cole, 1857); idem, *Creation, The Work of One Intelligence and Not the Product
of Physical Forces* (Ann Arbor: Young Men's Literary Association, 1858); idem, *Reconciliation
of Science and Religion* (New York: Harper & Brothers, 1877), 156-177.
[53] Alexander Winchell, *The Doctrine of Evolution: Its Data, Its Principles, Its Speculations,
and Its Theistic Bearings* (New York: Harper & Brothers, 1874), 8; idem, *Reconciliation of*

By the time Winchell's *Adamites and Pre-Adamites* appeared on
the scene in the spring of 1878 his evolutionary leanings were
already well known—so well known that the conventional suggestion that it was for his evolutionary partisanship that he lost
his Vanderbilt chair in 1878 seems questionable. Something had
ruffled Bishop Holland McTyeire, chairman of Vanderbilt's board
of trustees, and there was nothing like the whiff of a heresy that
taught that Adam was actually descended from black ancestors to
rub southern orthodoxy the wrong way! Winchell had brought
the full weight of his considerable scholarly prowess to bear on
the subject; he had accumulated a vast body of empirical data and
displayed a thorough grasp of the history of the preadamite theory. So, while providing a detailed exegesis of Peyrère's version,
he did not ignore the hints of preadamism detectable in the works
of figures such as Bory de Saint-Vincent and Bernard Hombron.[54]
Winchell consistently asserted the theological propriety of his
proposals; but in order to preserve his doctrinal credentials, at
least to his own satisfaction, he felt he had to make some major
structural alterations to traditional preadamism.

Thus from the outset he made it plain that his account—which
postulated that Adam was the natural offspring of Preadamite
parents—was thoroughly *monogenetic*. "I have not affirmed" he
categorically insisted, "—even like M'Causland and other ecclesiastical polygenisists—that mankind, one in moral nature, are
not one in origin; since I hold the blood of the first human stock
flows in the veins of every living being." By thus rupturing the
ties between preadamism and polygenism Winchell helped prepare the way for the later monogenetic version of preadamism
that would be deployed as a stratagem for meeting the challenge
of human evolution, even though he himself made the conscious
decision not to stir up that particular hornet's nest. "To assert that
man has advanced from the lowest human condition," he re-
minded his readers, "is not to assert that this condition was
reached by advance from the brute." And yet by calling attention
to the inferences that were not to be immediately drawn from
monogenetic preadamism, he was all the while raising in some
minds the possibility that those very implications, while not self-

*Science and Religion* (cit. n. 74), v. Cope's Lamarckianism is discussed in Peter J. Bowler,
"Edward Drinker Cope and the Changing Structure of Evolutionary Theory," *Isis,* 68
(1977): 249-265.
    [54] While Winchell did not specify his sources here (apart from Peyrère's treatise) it
seems likely that he had in mind J.B.G. Bory de Saint-Vincent, *L'Homme—Homo. Essai
Zoologique sur le Genre Humaine* (Paris, 1827); and Bernard Hombron, *Aventures les plus
Curieuses de Voyageurs,* 2 vols. (Paris, 1847).

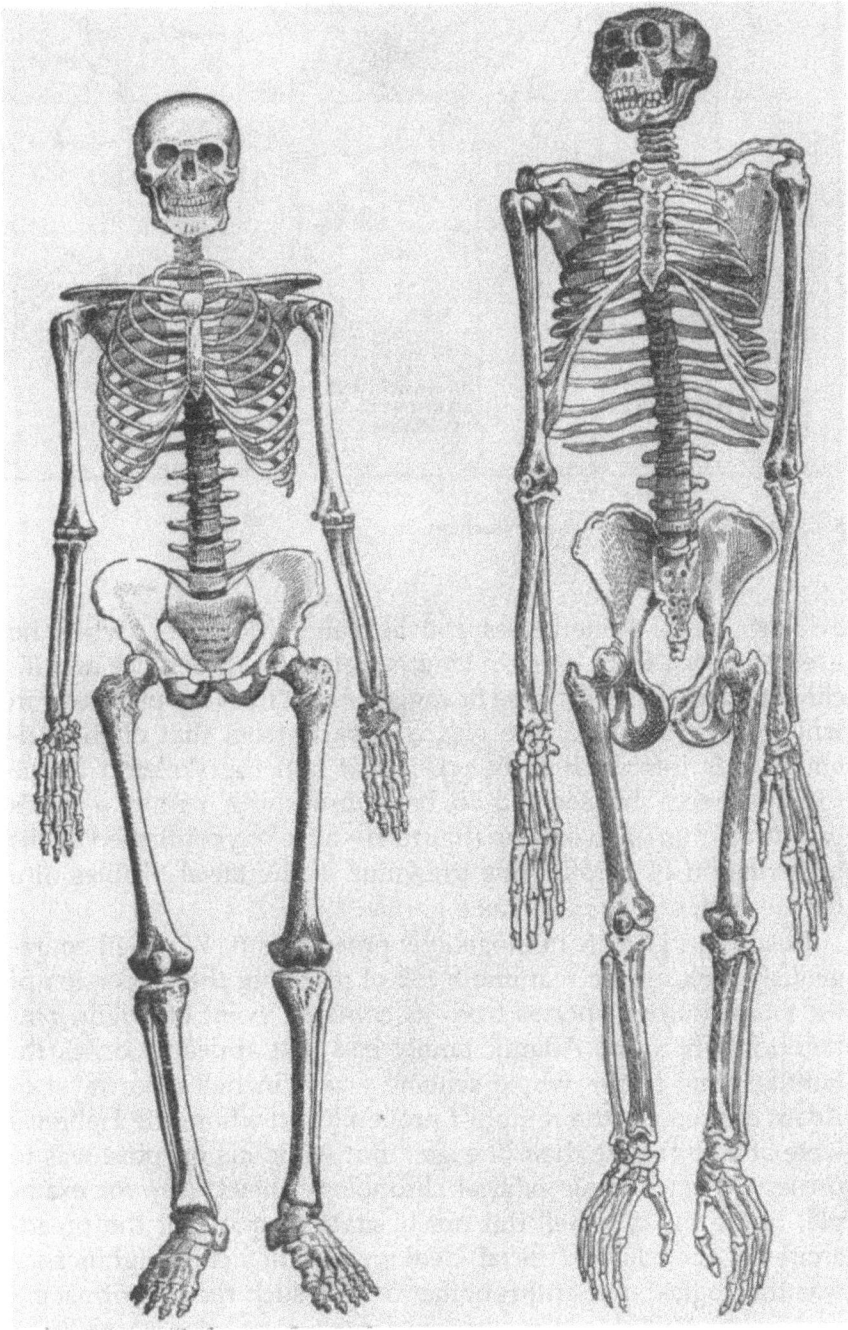

4. Skeletons of an Adamite (left) and a Chimpanzee (from Alexander Winchell, Pread-amites, 1880)

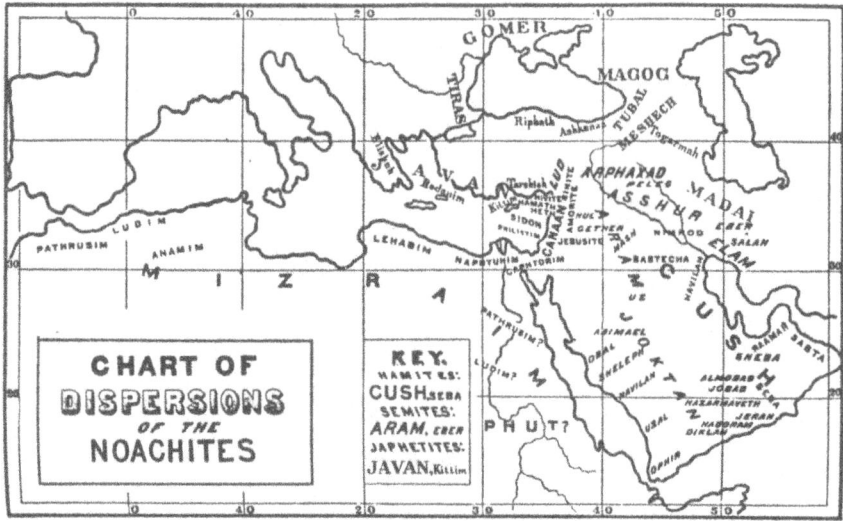

5. Chart of the Dispersions of the Noachites.

evident, might nonetheless still be valid. Moreover, when he used evidence from forearm length, pelvic circumference and inclination, and cephalic type to suggest that "in every particular in which the skeleton of the Negro departs from that of the Adamite, it is intermediate between that and the skeleton of the Chimpanzee," he seemed to be approaching within a hair's breadth of human evolution (figure 4)—a feeling confirmed by his presentation of parallel line drawings of the facial profiles of a female hottentot and a female gorilla.[55]

To substantiate his monogenetic preadamism, Winchell courageously took on the mammoth job of mapping the dispersion of the entire human species from its common point of origin, pinpointing where the Adamic family had first appeared on earth. Fundamental to the whole scheme was Winchell's portrayal of Adam as merely "the remotest progenitor to whom the Hebrews were able to retrace their lineage." But while his purpose was to preserve post-Adamic biblical chronology intact, (see for example, figure 5), Winchell did not hesitate to marshall the preadamites in the cause of racial ideology. Adam's chief significance was theological, not anthropometric; "no such racial contrast ex-

---

[55] Throughout I refer to the 1880 reprint edition: Alexander Winchell, *Preadamites; or a Demonstration of the Existence of Men Before Adam; Together with a Study of their Condition, Antiquity, Racial Affinities, and Progressive Dispersion over the Earth* (Chicago: S.C. Griggs and Company, 1880), v, 412, 249.

isted between the family of Adam and the nonadamites as to originate a racial repugnance" he insisted. Yet such racial compatibility applied only to Adam and the preadamic Dravida in his immediate genealogical line—the very group among whom Cain dwelt on his banishment from Eden. A quite definite racial repugnance existed between Adam and, for example, the Negritos or Hottentots (see figure 6, to be read from bottom upwards). Moreover, since racial differences were to be accounted for by natural means, not by divine creation, Winchell's preadamism, in contrast to earlier renditions of the theory, was thoroughly antagonistic to the notion of fixity of type.[56]

As might be expected, Winchell's scheme was just as racially serviceable as polygenetic preadamism (figure 7). Environmental modification of the human species had proceeded apace, to such a degree that there were now several physically, psychically, and linguistically distinct types of humanity. Moreover these differences, even if they were of environmental origin, were so deeply embedded in the history of the human species that the self-same racial types observable in the present had remained unchanged since ancient times, as drawings from Egyptian history attested (See figure 8). Besides, the anthropometric disparity between brachycephalic and dolichocephalic skulls supported his contention and, emboldened by such evidence, he drew up a comprehensive chart of world racial history which differentiated Negroid, Mongoloid, and Mediterranean types (Figure 6). Indeed his description of black inferiority, and the intensity of his miscegenation phobia were as rhetorically exuberant as any racial polygenist might exhibit. Advocates of racial intermixture were the special objects of his scorn, Wendell Phillips, Bishop Gilbert Haven and Canon George Rawlinson particularly coming under the whiplash of his tongue for their open advocacy of ethnic amalgamation. To Winchell the production of mulattoes was the blindest folly. Drawing on Stanford Hunt's somatic measurements of United States volunteers during the Civil War, he reported that the brain weight of the average pure Negro (1,331 grams)—so much lighter than that of whites (1,424 grams)—was nonetheless greater than one-quarter-white hybrids (1,319 grams). The implications for national racial policy were self-evident: the continued production of "freckled, blotched and mottled complexions, uncouth extravagances of features, short life, infecundity and general sanitary feebleness [which were] com-

---

[56] Ibid., 192.

MEDITERRANEANS

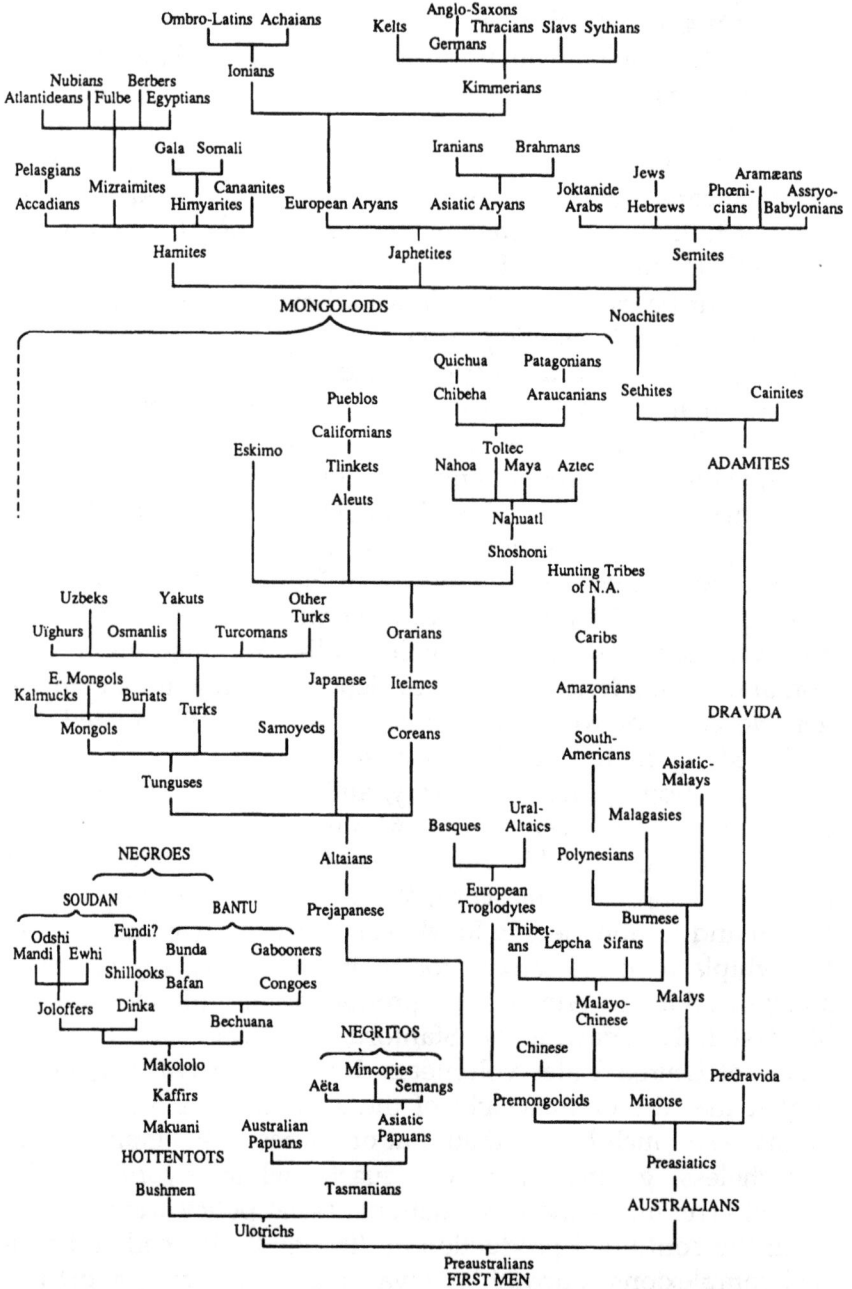

Ombro-Latins Achaians     Kelts   Anglo-Saxons   Thracians Slavs Sythians

Germans

Ionians

Nubians Berbers

Atlantideans | Fulbe | Egyptians

Kimmerians

Gala Somali      Iranians Brahmans

Pelasgians

Mizraimites   Canaanites     Joktanide   Jews    Aramæans

Accadians   Himyarites   European Aryans Asiatic Aryans   Arabs   Hebrews   Phœni- Assryo-

cians | Babylonians

Hamites      Japhetites      Semites

MONGOLOIDS      Noachites

Quichua Patagonians   Sethites    Cainites

Pueblos   Chibeha Araucanians

Californians        ADAMITES

Eskimo   Tlinkets   Nahoa | Maya Aztec   Toltec

Aleuts

Nahuatl

Shoshoni

Hunting Tribes of N.A.

Uzbeks Yakuts   Other Turks

Uïghurs Osmanlis Turcomans   Orarians   Caribs

E. Mongols       Japanese Itelmes   Amazonians

Kalmucks Buriats   Turks           DRAVIDA

Mongols   Samoyeds Coreans   South-Americans   Asiatic-Malays

Tunguses        Ural-Altaics   Malagasies

Basques      Polynesians

NEGROES     Altaians   European Troglodytes   Burmese

SOUDAN   BANTU Prejapanese   Thibet-ans Lepcha Sifans

Odshi Fundi?

Mandi | Ewhi   Bunda   Gabooners   Malayo-Chinese Malays

Shillooks Bafan Congoes

Joloffers Dinka   Bechuana   NEGRITOS   Chinese   Predravida

Makololo   Mincopies   Premongoloids Miaotse

Kaffirs   Aëta | Semangs

Makuani   Australian Papuans   Asiatic Papuans   Preasiatics

HOTTENTOTS

Bushmen   Tasmanians   AUSTRALIANS

Ulotrichs

Preaustralians
FIRST MEN

6. Winchell's Chart of World Racial History (redrawn from Alexander Winchell, Pread-amites, 1880)

7. Preadamites (from Alexander Winchell, Preadamites, 1880)

8. The Persistence of Racial Types (from Alexander Winchell, Preadamites, 1880)

mon characteristics of mulattoes" was both "shocking" to the "higher sentiments" and "destructive to the welfare of the nation and of humanity."[57]

Yet for all this unbridled racism, Winchell had committed in Southern eyes one unforgivable folly—he had made Adam the descendant of Blacks. On *this* point Winchell's thesis was similar to those of Prichard and M'Causland for he plainly confirmed that "Preadamism means simply that Adam is descended from a Black race, not the Black races from Adam." He even went so far as to suggest, reminiscent of Archibald Alexander, that the meaning of ADâMâH—"redness or ruddiness of colour"—favored the assumption that Adam himself was, if not pure black, nonetheless "strongly coloured." Still, if this *social* heterodoxy cost Winchell his Vanderbilt post, his motivations stemmed from his desire to remain *theologically* orthodox; for by keeping "the blood connection between the White and Black race intact," his proposals left the unity of mankind unimpaired.[58]

Winchell undoubtedly felt that his version of preadamism could relieve the doctrinal anxieties stirred by polygenism. Certain that other heresy hunters were waiting in the wings, he sought to forestall their charges too. Questions were sure to be raised about the soteriological status of the pre-and co-adamites, and he therefore argued that preadamism did "not interfere with current views of the catholic scope of the redemptive 'scheme'."[59] Here Winchell worked by analogy with the theological responses to the earlier challenge of the plurality of worlds,[60] citing men such as Chalmers, Brewster, and the Methodist Episcopal Bishop E.M. Marvin, who all extended redemptive privileges to other worlds.[61] Winchell's inference was immediate; if redemption

---

[57] Ibid., 83, 80, 81. Even Winchell's language was mild compared with that of Charles Carroll, *"The Negro a Beast" or "In the Image of God." The Reasoner of the Age, the Revelator of the Century: The Bible As It Is: The Negro and his Relation to the Human family. The Negro a Beast, but Created with Articulate Speech and Hands, that He may be of Service to His Master—the White Man. The Negro not the Son of Ham* (St. Louis: American Book and Bible House, 1900).

[58] Winchell, *Preadamites*, 285, 158, 285.

[59] Ibid., 286.

[60] In his splendid analysis, Brooke points out that, contrary perhaps to expectation, Scottish evangelicals remained open to the possibility of the plurality of worlds. See John Hedley Brooke, "Natural Theology and the Plurality of Worlds: Observations on the Brewster-Whewell Debate," *Annals of Science*, 34 (1977): 221-286. Also Paul Baxter, "Brewster, Evangelism and the Disruption of the Church of Scotland," in *"Martyr of Science". Sir David Brewster, 1781-1868*, ed. A.D. Morrison-Low and J.R.R. Christie (Edinburgh: Royal Scottish Museum, 1984), 45-50.

[61] Thus, Thomas Chalmers, *Astronomical Discourses* (Glasgow, 1817), Discourse IV; David Brewster, *More Worlds than One* (London, 1854); E.M. Marvin, *The Work of Christ* (St. Louis, 1867). Among other evangelical defenders of the idea of the plurality of worlds were Thomas Dick and Hugh Miller. See Thomas Dick, *The Sidereal Heavens* (London, 1840);

could encompass the farthest reaches of space, could it not reach back in time and thereby make "provision for the poor preadamites?"[62]

That Winchell found himself on the receiving end of Vanderbilt's evangelical disapproval does not mean that preadamism was entirely outlawed within the ranks of Methodist orthodoxy. The reception of his own work among the upper reaches of the denomination was altogether more sedate. In fact Daniel Whedon (1808-1885), one of American Methodists' most distinguished intellectual leaders for nearly a quarter of a century and editor of the prestigious *Methodist Quarterly Review*,[63] had already dispassionately reviewed M'Causland's volume, and Winchell did not hesitate to quote generously from Whedon's liberal-minded review in his own book. Whedon quickly saw how M'Causland's proposals could prevent "the violation of the sacred text" by safeguarding Old Testament chronology. As for soteriological objections, Whedon reported that "all evangelical theologians admit that the justifying power of Christ's death—had a retrospective effect," anonymously citing Peyrère's own argument that humanity's unity stemmed from moral identification with Adam, not lineal descent from him.[64] By allowing M'Causland's and indeed Peyrère's *polygenetic* Preadamism on to the stage of theological acceptability, Whedon thereby gave Winchell the opportunity of justifying his own monogenetic version as a good deal less open to charges of heresy.[65] In the long term, however, Whedon turned out to be a wavering ally and he ultimately drew back from wholeheartedly endorsing Winchell's proposals, perhaps because in the interim he had been favorably impressed with Edward Fontaine's *Ethnological Lectures on How the World was Peopled* (1872), which argued the case for the scientific legitimacy of pan-human derivation from an Edenic pair.[66] It was a bitter pill

---

idem, *The Christian Philosopher; or, The Connection of Science and Philosophy with Religion*, 2 vols. (London, 10th edition, 1846), vol. 1, pp. 302-304; Hugh Miller, "Geology Versus Astronomy" in *Essays*, (cit. n. 44), 364-379.

[62] Winchell, *Preadamites* (cit. n. 55), 290.

[63] See Emory Stevens Bucke (ed.), *The History of American Methodism. In Three volumes* (New York: Abingdon Press, 1964), vol. 2, pp. 189-190. Whedon regularly solicited articles from Winchell on scientific matters for the serial. See ibid., 385.

[64] [Daniel Whedon], Review of Adam and the Adamite by Dominick M'Causland, *Methodist Quarterly Review*, 53 (1871): 153-155.

[65] Winchell, *Preadamites*, (cit. n. 55), 287.

[66] [Daniel Whedon], Review of Adamites and Preadamites by Alexander Winchell, *Methodist Quarterly Review*, 60 (1878): 567; idem, Review of How the World was Peopled by Edward Fontaine, *Methodist Quarterly Review*, 54 (1872): 521-523. Winchell's project was also reviewed by Henry Colman, "Pre-adamites," *Methodist Review*, 7 (1891): 891-902. While Colman concluded that Winchell's case was "not proven," he did concede that "the

for Winchell to swallow, not least because he had already eulogized Whedon as "one of the noblest exponents of intelligent
theology" and "too shrewd to be fooled by the shriveled old ogre
of 'Orthodoxy,' who comes in the garb of Christianity, begging to
be defended from the assaults of common sense."[67]

This was not the first time Winchell had cause to feel betrayed
by fellow Methodists when he might have expected approbation.
Earlier he had prepared a "Preadamite" entry for M'Clintock and
Strong's *Cyclopaedia of Biblical, Theological, and Ecclesiastical Literature*. That the compilers allowed Winchell this space is itself significant, suggesting at least that Methodist debate on the subject
remained unimpeded. Still, the editors did not deliver the "Preadamite" entry over to Winchell *carte blanche*, for at key points
they appended editorial footnotes criticizing his arguments.
Some of their quibbles were relatively routine for example on
precise points of information—like the inadequacy of employing
certain astronomical data for dating Egyptian chronology. But
other complaints were rather more central. They resisted, for instance, what they took to be Winchell's efforts to force scripture
into harmony with science. Questioning the anthropological conclusions Winchell had drawn from the Hebrew meaning of the
term 'Adam', they hastily affirmed that the "statements of Scripture must stand or fall by themselves, when fairly expounded by
the usual laws of exegesis, and we are not at liberty to warp them
into accommodation with discoveries in other fields." Again,
when Winchell argued for Adam's descent from black ancestors
on the grounds of progress from inferior stock, they dissented
"*toto coelo*," especially from the "view that the Black races are in
any essential point inferior to the others." To them black degradation, a state taken for granted, stemmed from "unfavourable
surroundings . . . rather than . . . inherent lack of capacity." All
these editorial correctives were brought forward to preserve the
traditional Mosaic narrative. But they do reveal the complexity of
conceptual allegiances; the editors, biblical traditionalists as they
were, clearly displayed their distaste for racist sentiments,
whereas Winchell the scientific luminary based his re-reading of

theory of pre-Adamites conflicts with no biblical doctrine, and explains some otherwise
difficult Scripture texts. It is not so connected with Scripture as to become a theological
question, and must be decided by geology, ethnology, and history." Interestingly, Colman
described Peyrère (whom he called Peyrerius) as "an orthodox Dutch [sic] ecclesiastic."
ibid., 891.
[67] Winchell, *Preadamites*, (cit. n. 55), 288.

the biblical text on a thoroughly racist anthropology. The links between religious and social conservatism are, on this issue, far from clear cut.[68]

### Preadamism, Fundamentalism, and Anti-Evolutionism

Notwithstanding the Vanderbilt blot on Winchell's *curriculum vitae* and the tentativeness with which intellectual Wesleyanism approached his proposals, the preadamites were now well launched on their way to orthodoxy. What helped to establish their credibility among theological conservatives was their suitability for mediating between scripture and evolutionary theory. In point of fact future supporters of monogenetic preadamism, as we shall see, almost invariably recast the scheme in an evolutionary mold; almost, but not universally. So before turning to the evolutionary reworking of the theory, a word or two on the survival of Preadamism among twentieth century anti-evolutionists will be instructive.

One of the most remarkable, albeit terse, expressions of support among theological conservatives for the preadamite theory in its traditionally non-evolutionary guise is to be found in the writings of Reuben A. Torrey (1856-1928). Torrey was, in many respects, a fundamentalist *par excellence*, having served his time as one of D.L. Moody's foremost chiefs-of-staff.[69] Still, whatever the revivalist overtones of the movement he found himself spearheading through his editorial management of *The Fundamentals*, Torrey's intellectual roots in the cultural ethos of the New England tradition committed him to a firm belief in the mutual reinforcement of science and scripture. He accordingly applauded James Dana's concordist reading of Genesis, an understandable enthusiasm, perhaps, given that he had studied under Dana at Yale. Although he could not find it possible to negotiate a similar accommodation of scripture to Darwinian biology, he welcomed the preadamite as a peacemaker between biblical religion and archaeological science. Torrey claimed to have found the preadamite *within* the pages of Genesis, but he was delighted nonetheless that his biblical discovery should match scientific discoveries so well. Certainly his observations were, as I have indicated, exceedingly brief. Still, he did plainly state that "all verses after

---

[68] A[lexander] W[inchell], "Preadamite" in John M'Clintock and James Strong (eds.), *Cyclopaedia of Biblical, Theological, and Ecclesiastical Literature*, 10 volumes (New York: Arno Press, 1969; orig. New York: Harper & Brothers, 1877), vol. viii, pp. 484-492, on p. 486.

[69] See George M. Marsden, *Fundamentalism and American Culture. The Shaping of Twentieth-Century Evangelicalism 1870-1925* (New York: Oxford University Press, 1980), 47-48.

the first verse of Genesis I seem rather to refer to a refitting of the world that had been created, and had afterwards been plunged into chaos by the sin of some pre-Adamic race, to be the abode of the present race that inhabits it, the Adamic race." And again:

It should be said further that it may be that these ancient civilizations which are being discovered in the vicinity of Nineveh and elsewhere may be the remains of the pre-Adamic race already mentioned .. . . . No one need have the least fear of any discoveries that the archaeologists may make; for if it should be found that there were early civilizations thousands of years before Christ, it would not come into conflict whatever with what the Bible really teaches about the antiquity of man, the Adamic race.[70]

While Torrey's advocacy of preadamism may well seem surprising, given his fundamentalist theology, the presumption of antagonism within the fundamentalist coalition to science in general and to harmonizing schemes in particular is more a reflection of presuppositions on the part of present-day commentators than of the facts of the case. Indeed, the widespread adoption of the gap theory among conservative Protestants, from the Scottish Calvinist Thomas Chalmers to the American premillennialist C.I. Scofield may perhaps suggest why preadamism could linger long in that theological tradition. Earlier G.H. Palmer, for example, a vigorous advocate of Chalmers's gap thesis, produced his exceedingly popular *Earth's Earliest Ages* in 1876—a work that Scofield himself eulogized. And here he espoused the preadamite theory to account for certain aspects of the fossil record.[71] In the light of such alliances between the gap theory and the preadamite theory, one or two commentators have remarked that almost all gap theorists were advocates of some form of preadamism.[72]

A much more sustained defense of anti-evolutionary preadamism in the early twentieth century is to be found in the writings of Sir Ambrose Fleming FRS, president of the Victoria Institute, first president of the Evolution Protest Movement, and forty-one years' Professor of Electrical Technology at University College London.[73] If his presidency of the Victoria Institute com-

---

[70] R. A. Torrey, *Difficulties and Alleged Errors and Contradictions in the Bible* (London: James Nisbet, n.d., circa 1907), 31, 36.

[71] G.H. Pember, *Earth's Earliest Ages* (New York: Fleming H. Revell, 1876), 35.

[72] See Bernard Ramm, *The Christian View of Science and Scripture* (Exeter: Paternoster, 1971, first published 1954), 143; O.T. Allis, *God Spake by Moses. An Exposition of the Pentateuch* (Nutley, N.J.: Presbyterian and Reformed, 1947), 154-155.

[73] Biographical details are available in Sir Ambrose Fleming, *Memories of a Scientific Life* (London: Marshall, Morgan & Scott, n.d.); Charles Süsskind, "Fleming, John Ambrose," *Dictionary of Scientific Biography*, s.v. It should be pointed out that Fleming was not a creationist in the more recent sense of the term, for he was prepared to allow for some divinely-guided evolutionary change.

mitted Fleming "to investigate fully and impartially the most important questions of philosophy and science but more especially
those that bear upon the great truths revealed in holy scripture,
with a view to defending these truths against the oppositions of
science, falsely so-called,"[74] his association with the Evolution
Protest Movement was a public demonstration of his anti-evolutionary sentiments—the same as those expressed in his book-
length tract, *Evolution or Creation?*[75]

With his twin commitments to science and Christianity, it was
not surprising that Fleming would sooner or later turn to the
challenges to Christian theism arising from pre-historic archaeology and anthropology. He did so in his presidential address to the
Victoria Institute on Monday 14th January 1935. The *Daily Tele-
graph* gave prominent coverage to the lecture in its columns the
following day, and the anti-evolutionary sentiments that Sir Ambrose had expressed provoked comments from various quarters.
In the newspaper's Saturday edition he replied to his critics, and
here spelled out his own strategy for reconciliation, a strategy
subsequently elaborated in *The Origin of Mankind*. It was, fundamentally, the preadamite theory of M'Causland. Of course since
M'Causland's time new archaeological and anthropological findings had become available. In 1891 Eugene Dubois claimed to
have discovered fragments of *Pithecanthropus erectus* near Trinil in
Java and these were widely interpreted as the first hard evidence
for a primitive hominid. In 1907 Mauer Sands had unearthed part
of a jaw-bone with teeth of human type near Heidelberg and thus
was born *Homo heidelbergensis*. Most famous of all, however, was
Charles Dawson's 1912 discovery of human remains in a gravel
bed at Piltdown in Sussex that were subsequently pieced together
by Smith Woodward of the British Museum and Teilhard de Chardin. Though later shown in 1953 to have been fraudulent, these
remains were enthusiastically seized upon by evolutionists at the
time. Fleming was suspicious of all three and either argued, in
typically creationist fashion, that illustrators had let their imagi-
nations run wild by constructing both facial and skeletal profiles
from the scantiest of fragments, or generally questioned the ex-
travagant claims being made on the basis of negligible data.[76]

---

[74] These aims and objectives appeared on the inside cover of each issue of the Institute's
*Transactions*.

[75] Ambrose Fleming, *Evolution or Creation?* (London & Edinburgh: Marshall, Morgan &
Scott, n.d.).

[76] See Peter J. Bowler, *Evolution: The History of an Idea* (Berkeley: University of California
Press, 1984).

His approach to the evidence for "Neanderthal man" and the Cro-magnons was very different, however. Fleming fully acknowledged that the skull-cap and skeletal fragments of the former excavated at Düsseldorf in 1856, in 1887 near Spy in Belgium, and in the years up to 1914 in Krapina, Croatia, and southern France were compelling evidence. But he went on to suggest that this Neanderthal race was replaced in Europe by the superior Cro-magnons, complete skeletons of whom had been found in the Pyrenees and the Dordogne. The latter evidently possessed considerable ingenuity, artistic ability, and handicraft skills, as revealed in their bone and flint instruments and cave drawings. But while some archaeologists, like Schwalbe, argued that the Neanderthal evidence supported a human evolutionary conception, Fleming resisted such an interpretation. As he put it: "The upshot of it all is that we cannot arrange all the fossil remains of supposed 'man' in a lineal series gradually advancing in type or form from that of any anthropoid ape, or other mammal, up to the modern and now existing types of true man.[77] It was at this juncture, then, that the preadamite theory came to his rescue. Quite simply, Fleming suspected that the Neanderthals were a preadamic stock, whereas the specially created Cro-magnons were the Adamic antediluvians of the biblical narrative. That he should turn explicitly to M'Causland's polygenetic preadamism was thus doubly understandable for he was just as concerned as M'Causland to preserve a relatively literalistic hermeneutic of scripture. By this means Fleming could at once take seriously the patterns of early migration, crack those hoary exegetical chestnuts about Cain's wife and city, and yet again press the theory into the service of racial ideology.

And yet Fleming's preadamism displayed its own fair share of independent-mindedness. Whereas Adam's significance to traditional preadamism typically lay in his role as the progenitor of the Jewish race, Fleming built a more specifically "spiritual" component into his Adam, although accompanying physiological and psychological traits were certainly not overlooked. "The Creation of the Adamic man" he explained, "was the appearance on earth of a being more eminently endowed with psychical faculties of initiative, authority, and powers of intercommunication than before, and with special powers of intercourse with the Creator." In Fleming's mind, what confirmed his polygenetic preadamism

---

[77] Ambrose Fleming, *The Origin of Mankind Viewed from the Standpoint of Revelation and Research* (London & Edinburgh: Marshall, Morgan & Scott, n.d. [circa 1935]), 75.

was his belief that mankind could be divided into several distinct *species*: Caucasian, Mongoloid, and Negroid. So, by introducing a distinctively "spiritual" note (beyond the psychological) into his definition of the Adamic race, Fleming convinced himself that Caucasians possessed some *spiritual* superiority over other racial stocks. It was thus bad enough that inter-racial miscegenation produced physically degenerate offspring; that it produced spiritual mulattoes was too great to bear. From the beginning, it seemed, ethnic intermixture was also spiritual bastardy. In his own words:

If we then look at it from the point of view of the Biblical account of the Origin of Man, we see there recorded that the failure of the Adamic race to keep its privileged position originally given to it, and the inter-marriages of the descendants of the Adamite with prior created human beings brought about a state of moral degradation and violence tending to destroy the Divine purposes for that Adamic race. Hence came the Divine decision to destroy by a flood the bastard races, and begin again with a selected and God-obeying representative of the pure Adamic race.

And again:

. . . What it [the Noahite flood] did imply was a wiping out of all the bastard cross-breed race between Adamic and pre-Adamic man, and the beginning of the Caucasian race from the sons of Noah, Japheth, and Shem. It is allowable to presume that sufficient of the Mongolian, Negro, and other human species survived to continue the population of Eastern Asia and middle Africa.[78]

Fleming's resort to the preadamite theory was not, needless to say, universally welcomed. It certainly rubbed the anatomist and anthropologist Arthur Keith the wrong way. In point of fact it was largely because of Fleming's proposals that Keith put pen to paper to produce in March, 1935, a tract for the times entitled *Darwinism and Its Critics*, the first part of which had already appeared in the *Literary Guide*. Here Keith had Fleming clearly in his sights as he turned his big guns on those anti-Darwinians who still held to the "impossible theory" of special creation. His assault was wideranging, but in the present context his reaction to Fleming's preadamism is the main focus of concern:

Sir Ambrose ventures the opinion "that there have been pre-Adamitic races of

---

[78] Ibid., 115, 143-144. This was, of course, not the only way in which racism survived within evangelicalism. Another route was to emphasize that the black race was descended from Ham, one of Noah's sons, and was therefore the recipient of the curse in Genesis 9:25. Among evangelical attempts to refute this idea were George R. Horner, "Are the Negroes Cursed?" *His*, 7 (1947): 28-30; William A. Smalley and Marie Fetzer, "A Christian View of Anthropology," in American Scientific Affiliation, *Modern Science and Christian Faith. A Symposium on the Relationship of the Bible to Modern Science* (Wheaton, Ill.: Van Kampen Press, 2nd ed., enlarged, 1950), 114.

beings, whom I call hominoids in my address, but were not 'man' in the psychical and spiritual powers or possibilities in the Biblical sense of the word." He even ventures the opinion "that between true man and anthropoid apes there may have been some species of hominoids created." Is not Sir Ambrose taking an unwarranted liberty with the inspired word by introducing acts of creation and types of humanity of which there is no mention in the Mosaic record? Or would it not be more in keeping with scientific method to give up the theory of special creation, seeing that the truth has to be prosecuted to gain a verdict in its favour?[79]

Fleming's failure to convert Keith to his way of thinking hardly came as a surprise. He had more reason to be disappointed at the response he received from fellow evangelical Christians. His reviewer in the January, 1936, edition of *Christianity Today*, for example, found it "refreshing to find a scientist of Fleming's standing arguing from the facts of Christianity for not only the possibility but the probability that Man's origin is due to a special creative act of God." But there the praise ended. Fleming's suggestion "not only that Adam was not the first man but that he was the ancestor only of the Caucasians" was too much to swallow; and the reviewer went on to affirm the unity of the human race by calling upon the views of Benjamin B. Warfield—an important scholarly authority within conservative Protestantism. Ironically, as we shall see, Warfield himself had actually come to a transformed version of preadamism that was entirely in keeping with monogenism.[80] Again, in his *Systematic Theology* of 1939, Louis Berkhof, an American Calvinist theologian of Dutch birth, felt the need to reaffirm the standard Adamic narrative, and to present explicitly the views of Peyrère, Agassiz, Winchell, and particularly of Fleming as being utterly devoid of all scriptural support.[81]

Preadamism as an anthropo-theological thesis about human-racial origins, it is now clear, was a rather dynamic notion capable

---

[79] Sir Arthur Keith, *Darwinism and Its Critics* (London: Watts & Co., 1935), 20-21. The anti-evolutionism of Fleming's Victoria Institute address was also the subject of critical comment in the pages of *Nature* for 1935. See W. D. Lang, "Human Origin and Christian Doctrine," *Nature*, 136 (1935): 168-170. It is interesting in this context to draw attention to a later work published (I estimate) in the late 1940s by Anne Terry White entitled *Men Before Adam*. The writer here used the title "Men Before Adam" as a rhetorical tool to attack the creationism of William Jennings Bryan and to present in popular form the standard anthropological evidence for human evolution. The work was published by the Scientific Book Club, was thoroughly rationalist in spirit, and contained no reference to the biblical narrative. In fact the author concluded the volume by insisting that Sir Arthur Keith's own scientific vision, which she shared, needed the kind of popularization she believed her book could provide. The preadamite idea apparently could serve secular science in a variety of ways.

[80] Review of *The Origin of Mankind Viewed from the Standpoint of Revelation and Research* by Sir Ambrose Fleming, *Christianity Today*, 6, No. 6 (January 1936): 186.

[81] Louis Berkhof, *Systematic Theology* (London: Banner of Truth, 1971, first published 1939), 188-190.

of suiting a wide variety of social and scientific tastes. Some resorted to it as a means of subserving their own racial prejudices; others insisted that such extrapolations were entirely unwarranted. And while it was ordinarily equated with a polygenist stance it could be made compatible with monogenism. What is clear, however, is that depending on their particular reading of the preadamite theory, advocates found themselves committed to certain scientific theses about the differentiation of the human races. This further confirms that images of warfare or divorce between science and religion do little to further our understanding of those theories that were seen as genuine points of integration; for to supporters of preadamism (aside from mere polygenism) the belief that humans had existed before Adam was as much a *scientific* thesis as a *theological* construct.

Our perusal of the history of the preadamite theory to this point has thus shown it to be a rather elastic notion that was deployed initially to question the literal veracity of scripture and to advance skepticism in its various forms, and subsequently as a means of integrating a conservative hermeneutic of scripture with accounts of race history. With the widespread acceptance of evolutionary theory in the twentieth century, theological preadamism has yet again been transformed to meet this newest challenge. Accordingly, the survival of an evolutionary version of the thesis among theological conservatives should be our final concern.

## III. Evolutionary Preadamism and Conservative Theology

Whereas secular versions of preadamism could prevail in the early nineteenth century because talk of multiple creations provided a shared language among ethnologists and anthropologists, the same did not apply to the post-Darwinian scientific scene. Once the evolutionary theory with its naturalistic overtones had gained a firm foothold, the whole conception of preadamites only made sense to those who still attached some significance to the biblical Adam. There were indeed those, like Ernst Haeckel, who felt that the very idea of a first simply just had no scientific meaning. He himself was inclined towards monogenism, as he made clear in his *History of Creation*, but he explained that the question of monophyletic or polyphyletic descent, if construed as the search for the first human beings, was a fundamentally misconceived project. As he put it:

> . . . while we for many reasons believe that the different species of speechless primaeval men were all derived from a common ape-like form, we do not of course mean to say that *all men are descended from one pair*. This latter supposition, which our modern Indo-Germanic culture has taken from the Semitic myth of the Mosaic history of creation, is by no means tenable. The whole of the celebrated dispute, as to whether the human race is descended from a single pair or not, rests upon a completely false way of putting the question. It is just as senseless as the dispute as to whether all sporting dogs or all race-horses are descended from a single pair. . . . A "first human pair," or "a first man," has in fact never existed.[1]

Attributing mythological status to aspects of Old Testament theology, however, was not the prerogative of vituperative critics of religion like Haeckel or Vogt. There had long been theologians, particularly but not exclusively in Germany, who took a higher

---

[1] Ernst Haeckel, *The History of Creation: or the Development of the Earth and Its Inhabitants by the Action of Natural Causes. A Popular Exposition of the Doctrine of Evolution in General, and of that of Darwin, Goethe, and Lamarck in Particular*, translated and revised by E. Ray Lankester, 2 vols. (London: Kegan Paul, Trench & Co., 1883), vol. 2, p. 304. By the same token there were secular versions of human evolutionary theory that resonated with the ideology of preadamite racism. One case was the racial theorizing of the Australian geographer, Griffith Taylor, whose "Zones and Strata" theory involved the idea that the Negroes were closer to Neanderthals than to other modern humans. The idea was that the African races were the descendants of the earliest out-migrants from humanity's original cradleland and because they encountered a hostile environment they had retained many "primitive features." They were thus descended from inferior human forms. See T. Griffith Taylor, "The Zones and Strata Theory. A Biological Classification of Races," *Human Biology*, 8 (1936): 348-67.

critical view of scripture. For them, the creation story of Genesis with its depiction of Adam's creation, probation, and fall, were purely "mythological" and could therefore be sifted of all scientific content to leave behind the moral message as a kind of spiritual residue.[2]

In the light of these currents of scientific and religious thought, it is not surprising that the survival of the preadamite was left in the hands of conservative theologians, both Protestant and Catholic. Yet, as I have suggested, the theory that remained was transformed. In general terms the preadamite came to be regarded as a near-human or sub-human predecessor of the fully human Adam, the first being to be graced with the *imago dei*. As we shall see, such a conception could be interpreted in a variety of ways, but that it was fashioned into an evolutionary pattern is what most clearly demarcates it from earlier versions of the theory.

### Evolutionary Anthropology and Evangelical Theology

Because he occupied a "Chair of the Harmony of Science and Revealed Religion" specially created for him at Princeton in 1865, the first of its kind in any American college, the pronouncements of Charles Woodruff Shields (1825-1904) inevitably invite our perusal.[3] His institutional affiliation and his self-appointed task of surveying the entire gamut of scientific and theological learning in the immediate aftermath of the Darwinian onslaught together prepare us for a Presbyterian perspective on the question of human origins. Predictably, his monumental conspectus *Philosophia Ultima or Science of the Sciences* of 1888-1905, a work inviting comparison with the grand synthetic panoramas of Spencer or Fiske, took up the disparate claims of rational (scientific) anthropology as compared with those of revealed (scriptural) anthropology.

Shields was clearly familiar with the preadamite theory both in its role as an anthropological thesis and as a strategy for harmonizing science and scripture, and his references throughout the

---

[2] A useful entry point to the intellectual history of this subject is R.E. Clements, "The Study of the Old Testament," in *Nineteenth Century Religious Thought in the West*, 3 vols, ed. Ninian Smart, John Clayton, Steven Katz and Patrick Sherry (Cambridge: Cambridge University Press, 1985), vol. 3, pp. 109-141. See also Owen Chadwick, *The Secularization of the European Mind in the Nineteenth Century* (Cambridge: Cambridge University Press, 1975); Hans W. Frei, *The Eclipse of Biblical Narrative. A Study of Eighteenth and Nineteenth Century Hermeneutics* (New Haven: Yale University Press, 1974); J.W. Rogerson, *Old Testament Criticism in the Nineteenth Century: England and Germany* (London: SPCK, 1984).

[3] For brief biographical details, see "Shields, Charles Woodruff," in Appleton's *Cyclopaedia of American Biography*, ed. James Grant Wilson and John Fiske (New York: Appleton, 1888), s.v.; George McLean Harper, "Shields, Charles Woodruff," in *Dictionary of American Biography* (New York: Scribner's, 1935), s.v.

work to the writings of figures like Bruno, Peyrère, Kames, Agassiz, Nott, Gliddon, M'Causland, Poole, and a host of others, displayed his considerable erudition. At this juncture Shields's task was merely to lay before his public the range of scientific theories available, but it was already plain that he harbored some sympathy for the preadamite theory. Thus he was discomfited to find, for example, that "the guardians of orthodoxy in our day are denouncing Agassiz and [Edward] Forbes for a theory of co-adamite races, which might really support their own doctrine of a high Adamic covenant, as distinguished from mere inherited sin." By this assertion Shields intended to intimate that Adam could be conceived of as *representative* of the entire human species, and thereby relieve anthropology of the burden of commitment to universal descent from the Adamic pair. As he put it, "Whether the human family be of one race or many races, the first Adam and the second Adam would still be their chief moral representatives."[4] So it was no surprise that in 1900 he would clearly announce that the Genesis record dealt with the Adamic-Caucasian race as typical or representative of the entire human family and the bearer of divinely ordained salvation to the world—a view clearly assuming the continued survival of both pre- and co-Adamites across the reaches of history.[5]

Thus far, Shields's affirmations did not go beyond an endorsement of standard polygenetic preadamism. But there are hints that he was moving towards an evolutionary reworking of the concept. For instance, he had no sooner introduced Peyrère's system to his readers than he immediately went on to link it with Bishop Butler's treatment of "several questions of recent anthropology, such as the material origin of man, his development from an animal state, and his gradual predominance as the governing animal in our globe." And having made this connection, he presently added that earlier evangelical students of biblical anthropology such as Calvin and Beza were writing at a time when it "was too soon as yet to attempt any scientific verification of . . . dogmas, such as is beginning to be made, by associating co-Adamite and pre-Adamite theories of the savage and animal origin of man, with a special divine dispensation to Adam as the natural progenitor of the Caucasian race and federal representative of the

---

[4] Charles Woodruff Shields, *Philosophia Ultima or Science of the Sciences* (London: Samson Low, Marston, Searle, and Rivington, 1889, 3rd ed.), vol. 1, pp. 69, 370. Shields had already outlined these same options in *Religion and Science in their Relation to Philosophy. An Essay on the Present State of the Sciences* (New York: Scribner, Armstrong & Co., 1875).

[5] Charles Woodruff Shields, *The Scientific Evidences of Revealed Religion* (New York: Bishop Paddock Lectures for 1900), 124.

whole human family." But Shields could foresee a time, not far distant, when "the secular evolution of Adam from the animal species shall be claimed to be as scriptural and orthodox as that of the animal from the vegetable races, or that of the organized planet from the inorganic nebula." As it turned out, Shields was remarkably prescient on this very point, and his own resort to a covenant theology of the fall (which emphasized Adam's representativeness of the whole human species) as crucial to mediation between scripture and evolutionary anthropology was a move that later conservative theologians would make for precisely the same ends. Hence it is not surprising to find Shields himself affirming, in 1900, that if the "creation of man" was to be conceived of as "a process rather than as an act," it would be conceivable that "the animal organism of man" could have been endowed with "psychical qualities and divine resemblances . . . as new miraculous acts or subsequent processes during the historic period."[6]

If Shields sought an evangelical benediction on his evolutionary reading of the human story, he needed to look no farther than to the recently appointed Professor of Didactic and Polemic Theology in the Princeton Seminary, Benjamin Breckinridge Warfield, who took up that post in 1887. Warfield (1851-1921) was a Calvinist of solidly Old School convictions, the author of an architectonic defense of the fundamentalist doctrine of biblical inerrancy, and an immense authority among late nineteenth and twentieth century reformed evangelicals. That Warfield had found the conceptual resources in the long-established distinction between primary and secondary causes to meet the challenge of Darwinian biology was crucial to the story he wanted to tell about the origins of humanity. But before specifying the precise theological arrangements he devised, some reflections on his approach to the preadamite theory will be instructive.

In 1911, Warfield wrote an article for the *Princeton Theological Review* on the antiquity and unity of the human race. Because he felt that the mere age of the human race was irrelevant to theology he passed on to discuss theories that bore on the unity or diversity of the species, and to review some of the major scientific theories available. On turning to preadamism, he made a number of telling observations that throw much light on his own treatment of the subject. To him polygenism was to be equated with

---

[6] Shields, *Philosophia Ultima* (cit. n. 4), vol. 2, pp. 197, 203; *Scientific Evidences* (cit. n. 5), 118.

the *co*adamitism of writers like Paracelsus, and not with *pread-amism*, as was conventionally assumed. In his mind, "co-Adamit-ism [was] the attribution of the descent of the several chief racial types to separate original ancestors," whereas "pre-Adamitism . . . conceives man indeed as a single species, derived from one stock, but represents Adam not as the root of this stock, but as one of its products." The writings of Zanini, Isaac de la Peyrère, George Catlin, and Alexander Winchell, he reported, represented the monogenetic preadamite theory, while Gobineau, Nott, and Gliddon, Agassiz and Cordonière were advocates of polygenetic coadamism. Warfield himself warmed to neither of these ver-sions. On the one hand, he loathed polygenism in every shape and form, as well as the racial pride that typically went with it, and on the other, he believed that Christianity's theological struc-ture was bound up with Adam as the father of all humanity, not just of the Jews. Yet he did make reference to another "sort of pre-Adamitism [that] has continued to be taught by a series of philosophical speculators from Schelling down, which looks upon Adam as the first real man, rising in developed humanity above the low, beastlike condition of his ancestors."There Warf-ield left the matter, but his insistence that monogenism was "a necessary corollary of the evolutionary hypothesis," I would sug-gest, already raised the suspicion that this other "sort" of Pread-amism might have some validity.[7]

---

[7] B.B. Warfield, "On the Antiquity and the Unity of the Human Race," *Princeton Theo-logical Review*, 9 (1911): 1-25, reprinted in *Biblical and Theological Studies* (Philadelphia: Presbyterian and Reformed Publishing Co., 1968), 238-261, on p. 256. Warfield's enthusi-asm for Shields's efforts is apparent in his review of *Philosophia Ultima* in *Princeton Theo-logical Review* 4 (1906): 541-542. In the Netherlands, precisely the same distinctions were made by Herman Bavinck. "Na de renaissance kwam hier en daar het denkbeeld weer op van een verschillenden oorsprong van het menschelijk geslacht. Dit denkbeeld trad nu eens op in den vorm van eigenlijk polygenisme bij Caesalpinus, bij Blount en andere deïsten; deels als coadamitisme, d.i. afstamming der verschillende rassen van verschil-lende stamvaders, bij Paracelsus e.a.; deels als praeadamitisme, d.i. afstamming der wilde en donkerkleurige volken van een stamvader vóór Adam, terwijl deze dan alleen de stamvader was van de Joden of ook van de blanke menscheid, bij Zanini en vooral bij Isaac de la Peyrère. . . . Een ander soort van polygenisme werd door Schelling geleerd. Hij nam ook vele rassen van menschen aan vóór Adam, maar deze hadden zich van een laag, dierlijk standpunt zoo opgeheven en ontwikkeld, dat ze eindelijk hem voortbrachten, in wien het menschelijke eerst tot openbaring kwam en die daarom terecht den naam van den mensch, haadam, dragen kon. En evenzoo werd een zeker praeadamitisme geleerd door Oken, Carus, Baumgartner, Perty, Bunsen." [After the Renaissance the idea of dif-ferent origins of the human race reemerged in several places. This idea appeared in different forms: at times as a real polygenism for Caesalpinus, Blount and other deists; as co-adamites, i.e. the descent of different races from different ancestors, by Paracelsus and others; as pre-adamites, i.e. descent of the savage and dark colored peoples from an ancestor before Adam, who was the sole ancestor of the Jews and the white race, by Zanini and especially for Isaac de la Peyrère. . . . A different type of polygenism was propagated by Schelling. He assumed that many races of humans had existed before Adam; these

These inklings are not confirmed until we turn to Warfield's own views on the emergence of the human species. In an exposition of "Calvin's Doctrine of the Creation" in 1915, he made the case that Calvin had restricted the term "creation" to the initial act of *creation ex nihilo*. The only exception to this was the creation of the soul, for Calvin held to the Creationist, as opposed to Traducianist view, that every human soul throughout the history of propagation was an immediate, not mediate, creation. Accordingly, since all other modifications of the primeval "indigested mass" was by "means of the interaction of its intrinsic forces," Warfield could claim that Calvin's was a "very pure evolutionary scheme." So far as the human species was concerned, it left open the possibility that the human body might have undergone a long history of evolutionary modification prior to receiving, by an act of creation, a soul. Thus, when Warfield came to reviewing James Orr's *God's Image in Man*, he allowed the possibility that the human body had been formed in emergent evolutionary fashion, "at a leap from brutish parents," and then fitted with a "truly human soul." The compatibility of this conception with the alternative preadamite scheme that Warfield himself had identified is clearly apparent.[8]

If Warfield left evolutionary preadamism as no more than an inference to be drawn from his writings, other Protestant conservatives were far more explicit in their endorsement of this new version of the theory. Take, for example, A. Rendle Short (1880-1953), the Royal College of Surgeons, Hunterian Professor. Short was a distinguished surgeon and vigorous apologist for evangelical Christianity. Accordingly, in the latter capacity, he took up the subject of the "Problem of Man's Origin" in a volume on science and the Bible that appeared during the 1930s. Here he hinted at the possibility that there "might conceivably have been

---

races, however, had undergone such development and "elevation" from an animal standpoint that they finally brought forth Adam in whom human characteristics were first revealed, and therefore he could justifiably carry the name of the human, haadam. Similarly, Oken, Carus, Baumgartner, Perty, and Bunsen taught a certain preadamism.] H. Bavinck, *Gereformeerde Dogmatiek* (Kampen: J.H. Kok, 1918), 558-60. Bavinck resolutely rejected polygenetic preadamism, and while he does not give a final verdict on the evolutionary reworking of the theory one can sense a certain unease with even the most monogenetic versions of preadamism.

[8] B.B. Warfield, "Calvin's Doctrine of the Creation," *Princeton Theological Review* 13 (1915): 190-255, on pp. 208, 209; idem, *Review of God's Image in Man* by James Orr, *Princeton Theological Review*, 4 (1906): 555-558. It should be pointed out that Warfield's interpretation of Calvin on this point has been rejected by John Murray, "Calvin's Doctrine of Creation," *Westminster Theological Journal*, 17 (1954): 21-43; and by Richard Stauffer, "L'Exégèse de Genèse 1, 1-3 chez Luther et Calvin," In *Principio: Interprétation des Premiers Versets de la Genèse* (Paris: Etudes Augustiennes, 1973), 290.

pre-Adamite creatures with the body and mind of a man, but not the spirit and capacity for God and eternity." Certainly at this stage, he concluded that all this was "very difficult and speculative;" but by 1942, when he produced *Modern Discovery and the Bible*—a work destined to remain popular within British evangelicalism—he was rather more enthusiastic. Now he pointed out that it was by no means certain that *Homo Neanderthalensis* was "to be regarded as man in the Bible sense of the word." Rather the Neanderthals could well be preadamite. Indeed it was just possible that the biblical Adam could have been formed by the infusion of spiritual qualities into some preadamic creature. As he explained:

They [the Neanderthals] may have been pre-Adamic, and Adam verily was a new creation, with spiritual qualities that they lacked. What sort of material the Creator used to make man, whether the dust of the earth directly, or the pre-existing body of a beast, we leave an open question.[9]

While Short thus allowed the possibility of human evolution, he was far from happy with the conventional Darwinian explanation, and found himself much more inclined towards the mutation theory of Lev Simonovich Berg, the Russian geographer and ichthyologist. Accordingly he spelled out in some detail the evidence that Berg had accumulated in support of what he called "nomogenesis." Its significance in the present context is simply that it kept open the possibility of an evolutionary jump from preadamite to adamite—precisely the sort of arrangement that Warfield had envisaged.[10]

Short's diagnosis was to cast a long shadow on British evangelical thought. In the standard conservative commentary on the book of Genesis that appeared in 1967, for example, Derek Kidner explicitly returned to Short's preadamite theory to make his peace with evolution. He suggested that if, as in all likelihood, "God initially shaped man by a process of evolution, it would follow that a considerable stock of near-humans preceded the first true man, and it would be arbitrary to picture these as mindless brutes." On the contrary, by disengaging the idea of the image of God from the notion of "rationality," Kidner was able to conceive of culturally sophisticated preadamites who were "of comparable intelligence" to Adam, but still lacking the spiritual *imago dei*.

---

[9] A. Rendle Short, *The Bible and Modern Research* (London & Edinburgh: Marshall, Morgan & Scott, n.d.), 57; idem, *Modern Discovery and the Bible* (London: Inter-Varsity, 1961, first pub. 1942), 114.

[10] Short, *Modern Discovery*, 49-53, 65, 101, 115. Information on Berg is available in V.V. Tikhomirov, "Berg, Lev Simonovich," in *Dictionary of Scientific Biography*, s.v.

More, he went so far as to suggest that after the "creation" of Adam, "God may now have conferred His image on Adam's collaterals, to bring them into the same realm of being." And by conceiving of Adam as the "federal" rather than genetic head of humanity, the knotty problem of original sin could be disentangled.[11]

In broad terms, this self-same model has been adopted by the evangelical statesman John R. W. Stott and the geneticist R.J. Berry, both of whom speak of preadamic hominids with very substantial cultural acquisitions and intellectual prowess. To them, the image of God is again given a theological, rather than psychical or physiological, meaning, thereby leaving room for the evolution of *Homo sapiens* prior to its transformation into *Homo divinus*, the biblical Adam.[12] Meanwhile, in the interim, Bernard Ramm had kept the preadamite theory before the American evangelical mind as a means of retaining anthropology within the confines of biblical orthodoxy. Ramm's *Christian View of Science and Scripture* came to achieve almost legendary status among those orthodox Protestants who wanted to keep an open mind on the scientific enterprise. And not without good cause. His survey of the relevant literature was remarkable; on the preadamite option alone he traced its origins back to Peyrère, referenced its manifestation among fundamentalists like Pember and Torrey, and spoke of its interpretative versatility. He himself felt that the theory had "vexing problems" but his review clearly substantiated its viability as an evangelical option.[13]

To these theological conservatives, preadamism in its evolutionary guise has kept alive the marriage of science and religion.

---

[11] Derek Kidner, *Genesis. An Introduction and Commentary* (London: Tyndale, 1967), 28-29.

[12] John R. W. Stott, *Understanding the Bible. The Story of the Old Testament* (London: Scripture Union, 1978), 5; R.J. Berry, *Adam and the Ape. A Christian Approach to the Theory of Evolution* (London: Falcon, 1975), 44. Berry, it should be pointed out, specifically uses Kidner's federal theology as a means of preserving human spiritual unity without commitment to physical descent of all humankind from Adam. A new version of this volume has been published as *God and Evolution* (London: Hodder, 1988). In the early 1960s Russell Mixter's similar belief that prior to 10-15,000 B.P. "there were men as judged from their anatomy (upright posture) but . . . they were pre-Adamic creatures and not scriptural man" lead to a controversy at Wheaton College, due to the protests of a certain R.T. Ketcham, National Consultant of the General Association of Regular Baptist Churches. Mixter's statement comes from a letter he sent to Ketcham, November 20, 1962, and is cited with the permission of the author. I am grateful to Ronald Numbers of the University of Wisconsin-Madison for drawing my attention to this episode.

[13] Ramm, *Christian View of Science and Scripture* (Exeter: Paternoster, 1971, first published 1954), 222. Ramm's significance within the circle of evangelicals in science may be gauged from the Bernard Ramm Festscrift which appeared as a special issue of the *Journal of the American Scientific Affiliation*, 31, number 4 (1979).

For others, it has played an even greater role, substantially shaping exegesis of the Genesis narrative. This is so, for example, in the case of E.K. Victor Pearce, whose book *Who was Adam?* was specifically written to substantiate preadamic claims from a detailed reading of the text as much as from the field evidence of prehistoric anthropology and archaeology. With a clear commitment to biblical inerrancy, Pearce set out to square the claims of science with scripture in a manner worthy of the grand concordist schemes of the nineteenth century harmonizing geologists Miller, Guyot, and Silliman. Pearce's claim is simply that the early chapters of Genesis house two distinct creation stories, the former focusing on the creation of the human species, the latter narrating the life-story of an individual, Adam. As evidence for the existence of the former preadamites, Pearce painstakingly correlated specific biblical phrases with particular archaeological artefacts from prehistory, and thereby showed how preadamism could be incorporated into the most literal reading of the Genesis chronicle.[14]

None of the foregoing must be taken to imply that evolutionary preadamism has as yet won general evangelical approval.[15] Latter-day creationists, for instance, who brook no tampering with the special creation of Adam from the dust of the ground, make the preadamite as unwelcome as those liberals for whom Adam has nothing but mythological significance. It is not surprising, then, that beyond orthodox protestantism, preadamism still survives in the writings of various Catholic theologians. Indeed, the standard Catholic encyclopedias of the twentieth century have continued to include entries on the Preadamites that invariably review and still refute Peyrère's theological rendition of preadamism.

---

[14] R.K. Victor Pearce, *Who was Adam?* (Exeter: Paternoster, 1969). On the basis of the words in Gen. 2.5, for example, Pearce claims that he finds there depicted "the tundra vegetation and dry outward-blowing wind conditions," and from Gen. 2.8 "a clue to how Adam might have discovered the secret of cross-pollinating wild grasses to produce cereals." ibid., 52, 54.

[15] Thus the Dutch Reformed zoologist, Jan Lever, found scientific and theological difficulties with the various evolutionary versions of preadamism. See Jan Lever, *Creation and Evolution* translated by Peter G. Berkhout (Grand Rapids: Grand Rapids International Publications, 1958), 172. William Smalley (a Missionary) and Marie Fetzer (a Wheaton College instructor in anthropology) preferred to opt for an early date for Adam due to difficulties they had with preadamism. See Smalley and Fetzer, "A Christian View of Anthropology," in American Scientific Affiliation, *Modern Science and Christian Faith. A Symposium on the Relationship of the Bible to Modern Science* (Wheaton, Ill.: Van Kampen Press, 2nd ed., enlarged, 1950), 186.

## Preadamism and Modern Catholic Theology

On at least one element in the nexus of preadamite ideas the
Catholic tradition has remained uncompromisingly hostile,
namely, the belief in the continued existence of preadamites
alongside the adamic family. This version of coadamism was
branded as heretical by Pesch in 1910, while the papal encyclical
of 1950, *Humani Generis*, insisted that "Christ's faithful cannot
embrace the opinion that after Adam there existed on this earth
true men who did not take their origin through natural generation
from him."[16] Yet in 1913 A.J. Mass did allow for the possibility of
*preadamism* when he asserted that "the existence of a human race
(or human races) extinct before the time of Adam . . . is as little
connected with the truth of our revealed dogmas as the question
whether one or more of the stars are inhabited by rational beings
resembling man," and called on the supporting testimony of
Catholic dogmaticians such as Palmieri and Fabre d'Envieu. Although Mass did not outlaw preadamism, he felt that neither the
scientific evidence nor Peyrère's theological system were compelling, nor was what he called the "political-social Preadamism" of
M'Causland, Poole, and the American polygenetic ethnologists.[17]
The monogenist slant of Catholic thought, of course, needs to be
interpreted with caution. For one thing, monogenism is entirely
compatible, as we have seen, with certain versions of preadamism, and, as if to confirm that very point, O.W. Garrigan noted
in 1967 that for Catholics "any preadamite hypothesis must allow
in some way for the doctrine of original sin and the unity of the
present human race."[18] The failure of *Humani Generis*, at least in
the eyes of some commentators, to make a firm statement on the
issue of preadamism (aside from coadamism) left room for con-

---

[16] Quoted in O.W. Garrigan, "Preadamites," in *New Catholic Encyclopedia* (New York:
McGraw-Hill, 1967), vol. 11, p. 702. Clearly the idea had been flirted with by late nineteenth century Catholics. Thus, N. Joly, himself a Catholic, cited a certain Abbé Fabre as
asserting that "prehistoric archaeology and palaeontology may, without running counter
to the Scriptures, discover in the tertiary beds and in those of the early part of the
quaternary the traces of *pre-Adamites*." [Emphasis in original]. N. Joly, *Man before Metals*
(New York: Appleton, 1889), 186.

[17] A.J. Mass, "Preadamites," in *The Catholic Encyclopedia. An International Work of Refer-
ence on the Constitution, Doctrine, Discipline, and History of the Catholic Church* (New York:
Encyclopedia Press, 1913), vol. 12, pp. 370-371. See also E. Amann, "Préadamites," in
*Dictionnaire de Théologie Catholique contenant l'Exposé des Doctrines de la Théologie Catholique.
Leurs Preuves et Leur Histoire* (Paris: Librairie Letouzey, 1935), vol. 6, pp. 2793-2800.

[18] Garrigan, "Preadamites," (cit. n. 16).

siderable debate on the subject, particularly among French and Italian Catholics.[19]

All this suggests that the Catholic confession of the unity of the human race is no more monolithic than its Protestant counterpart. Take, for example, the hearty defense of monogenism by Humphrey Johnson during the inter-war years. He pointed out that it was once fashionable to "suggest that neolithic man was 'adamite' and paléolithic man 'pre-adamite'." But such a maneuver he felt was gratuitous. "Provided that we do not demand any high degree of mental development in the first true man," he affirmed, "there is . . . no very grave difficulty in deriving all rational beings who have ever inhabited our planet, from one pair."[20] Thus Johnson's rejection of polygenetic preadamism was orchestrated by pushing the date of Adam's first appearance farther and farther back in time. In order to sustain his reading of race history, moreover, he had to couple his monogenetic stance with a belief in retrogression. In fact, he argued in 1950 that the discovery in Palestine and Moraira of human remains exhibiting a combination of modern and Neanderthaloid features suggested that the two racial types were less divergent than at first believed; more, "the fact that those Neanderthal skulls in which the pithecoid traits are most accentuated are chronologically the most recent [was] an indication that retrogression has been at work."[21] Such a stance, of course, had ideological possibilities, particularly when linked to the outmoded theory of ontogeny recapitulating phylogeny. So, for all his commitment to the organic unity of the race, he openly declared that since "the history of the individual

---

[19] For example, F. Asensio, "De Persona Adae et de Peccato Originali Originante Secundum Genesim," *Gregorianum*, 29 (1948): 464-526; T. Ayuso Marazuela, "Poligenismo y Evoluzionismo a la Luz de la Biblia y de la Teología," *Arbor*, 19 (1951): 347-372; J. Havet, "L'Encyclique "Humani Generis" et le Polygénisme," *Revue Diocesaine de Namur*, 6 (1951): 114-127; Guy Picard, "La Science Expérimentale est-elle Favorable au Polygénisme?" *Sciences Ecclesiastiques*, 4 (1951): 65-89; J. Bataini, "Monogénisme et Polygénisme. Une Explication Hybride," *Divus Thomas* (Piac.), 30 (1953): 363-369; P. Denis, *Les Origines du Monde et de l'Humanité* (Liège, 1950); M. García Cordero, "Evolucionismo, Poligenismo y Exegesis Biblica," *Ciencia* 78 (1951): 465-475, 477- 479; M.M. Labourdette, *Le Péché Originel et les Origines de l'Homme* (Paris: Alsatia, 1953); G. Colombo, "Transformismo Antropologico e teologia," *Scuola Cattolica*, 77 (1949): 17-43; F. Ceuppens, "Le Polygénisme et la Bible," *Angelicum*, 24 (1947): 20-32. These works are a small sample of a larger body of Catholic literature on the subject. A more comprehensive listing is to be found in Karl Rahner, *Theological Investigations. Volume 1. God, Christ, Mary and Grace* (London: Darton, Longman & Todd, 1961, first published in German in 1954), 229-230.

[20] Humphrey Johnson, "The Problem of Prehistoric Man," *The Tablet*, 1939, August 12, 211.

[21] Quoted in C. Lattey, "The Encyclical "Humani Generis" . . .," *Scripture*, 4 (1951): 278-279 on p. 279.

recapitulates the history of race we may infer that the backward races of mankind, such as the Negro and the Australian, are to some extent mental degenerates."[22]

Although Johnson's monogenist anthropology ruled out preadamism, other Catholic statements approached the evolutionary version adopted among Protestant theologians. A convenient point of access to the twentieth century debate among Catholics on the issue is centered on the contributions of Ernest C. Messenger. Even before he had made his personal thoughts on the incorporation of evolutionary thought into Catholic theology available to his contemporaries, Messenger had already done much to disseminate his aims among English-speaking Catholics by his translation of Henry de Dorlodot's *Darwinism and Catholic Thought* in 1925. Dorlodot was director of the Geological Institute at Louvain University, and much of his work was understandably devoted to an elucidation of the writings in the classical Catholic tradition in order to substantiate their compatibility with evolution. Accordingly Dorlodot perused and commented on the relevant encyclicals, the biblical commission of Leo XIII, and the exegesis of the Hexameron, as well as the writings of the Fathers and the Aristotelian scholastics, and showed to his own satisfaction that they cohered with a natural evolutionary account of species transformation. Throughout his work, Dorlodot relied on the time-honored distinction between primary and secondary causes. As for the question of human evolution, Dorlodot resorted to Catholic principles of embryological development and cited St. Thomas Aquinas to substantiate the view that "the human embryo before it reaches a state fitted to be animated by the intellectual soul must have been animated successively, not simply by a vegetative soul and then by an animal soul—but rather must have passed—*per multas generationes et corruptiones*—through a great number of substantial transformations."[23] On the basis of embryology, then, as we shall again note, an evolutionary account of race history might be erected. Indeed for Dorlodot the only break in the chain of natural evolution was the creation of the human soul—a view which we have already encountered within Protestant theology.

Ernest Messenger's own *Evolution and Theology* understandably drew on Dorlodot's historical groundwork, so much so that one reviewer quipped that the best parts of his book were those based

---

[22] H.J.T. Johnson, *Anthropology and the Fall* (Oxford; Blackwell, 1923), 51.

[23] Dorlodot, *Darwinism and Catholic Thought*, translated by E.C. Messenger (London: Burns, Oates & Co., 1922), 109-110.

on Dorlodot's analysis.[24] Here again much space was devoted to the elucidation of key Catholic sources—St. Ephrem, St. Basil, St. Gregory of Nyssa, St. Ambrose and so on—all to support the legitimacy of the thesis that the creation of species was by natural causes. Then he moved on to a much more detailed examination of the "the origin of man," and argued that the emergence of the human race by way of phylogenetic descent was entirely compatible with tradition, and that opposition from modern theologians stemmed from their overly literalistic reading of the Old Testament. At this point the preadamite entered the arena, and once established began the task of mediating between evolution and Catholic dogma. Messenger's argument was that once the possibility of preadamite human forms was entertained, the need for postulating a creationist account of Adam's physical form withered away; besides, the preadamites had an independent value "in accounting for the many apparently imperfect types of humanity which recent archaeology has revealed." So, if the whole purpose of creation was to "lead up to man," it was "in every way fitting that God should thus have made use of secondary causes in the formation of Adam's body," and this in turn implied that other "creatures may have taken an active part in preparing for his coming."[25]

Messenger's work received "fairly wide distribution" in the years immediately after 1931, but stock copies of the book were destroyed in air raids on England during the Second World War.[26] Still, it certainly had provoked a spate of comment in the Catholic press; and so, in 1949, Messenger brought out a collection of these essays under the title *Theology and Evolution*. Many of his reviewers painstakingly interrogated Catholic tradition to determine the legitimacy of Messenger's reading, and many of the arguments and counter-arguments addressed subjects no less central to evangelical Protestant theology. Bernard Ramm, for example, later commended the volume as an exemplary piece of theological probing, "a masterpiece of erudition" as he called it, that the evangelical tradition would do well to emulate.[27] The nuances of the debate need not detain us further, save to note

---

[24] Abbé J. Gross, "The Problem of Origins in recent Theology," *Theology and Evolution (A Sequel to Evolution and Theology)* ed. E.C. Messenger (London: Sand & Co., 1949), 124-145, on p. 130. Translated from *Revue des Sciences Religieuses* for January 1933.

[25] Messenger, *Evolution and Theology* (London: Burns, Oates and Washburne, 1931), 277, 279.

[26] Messenger, "Introduction" in *Theology and Evolution* ed. Messenger (cit. n. 24), 1.

[27] Ramm, *Christian View of Science and Scripture* (Exeter: Paternoster, 1971, first published 1954), 225.

that when P.G.M. Rhodes, a sympathetic commentator, observed
that it "may well be imagined that God so directed the course of
evolution that animals which might be described as sub-human
came into existence, possessing nearly, but not quite, the human
configuration, without the rational soul," this was essentially the
evolutionary preadamism that we have already scrutinized.[28]

Although evangelical and Catholic observers often shared similar doctrinal strategies in their encounter with human evolutionary theory, on one particular issue a noticeable divergence is
apparent, namely, the use of embryology. Thus, due to its ana-
logical value, embryological matters came to the surface in Messenger's volume. He himself presented a thumb-nail sketch of the
history of embryology and reprinted an address he had given on
Aquinas's embryological thought, in order to vindicate the mediate animation theory that Dorlodot had championed. The argument here revolved around the question of when the soul entered
the human foetus. Messenger and Dorlodot both urged that the
idea of its infusion only *after* a certain period of development
(hence *mediate* rather than *immediate* animation) was thoroughly
in keeping with tradition. The discussion soon moved off into
questions of technical Catholic dogmatics, but its significance to
the preadamite question stemmed from its bearing on the issue of
recapitulation. Messenger felt that Haeckel's "Law of Recapitulation," as originally formulated, had little to commend it. "But
viewed in the light of the Thomist theory of the succession of
forms in the human embryo . . . it is highly suggestive," he went
on. "For, if a human being at the present time goes through first
a vegetative and then a sensitive or animal stage, it is difficult not
to think it likely that the human race as a whole may have had a
similar history." In his mind, therefore, advocates of the immediate animation theory, like W. McGarry, generally opposed human evolution, while defenders of the mediate theory welcomed
it.[29]

Because it was the first document to take up the question of

---

[28] P.G.M. Rhodes, "The Problem of Man's Origin," in *Theology and Evolution* ed. Messenger (cit. n. 24), 3-9, on p. 9.
[29] E.C. Messenger, "Evolution and Theology To-Day: A Re-Examination of the Problems," in *Theology and Evolution* ed. Messenger (cit. n. 126), pp. 172-216, on pp. 194-195. In
this same volume Messenger also wrote the following articles: "Outline of Embryology, in
the Light of Modern Science" (ibid., 221-232); "A Short History of Embryology," (ibid.,
233-242); and "The Embryology of St. Thomas Aquinas" (ibid., 243-258). The volume also
included the following pieces by Henry de Dorlodot, "A Vindication of the Mediate Animation Theory" (ibid., 259-283); "An Objection from Moral Theology: The Question of
Abortion and the Mediate Animation Theory," (ibid., 301-312); "A Formal Answer to
Objections Against the Mediate Animation Theory," (ibid., 313-326).

human origins, the appearance of the encyclical *Humani Generis* in 1950 constitutes an important moment in the history of pronouncements from Catholic officialdom on the topic. The canon was prepared in view of the denial of monogenism by some ethnologists, particularly in France, and occasioned widespread comment in the Catholic media.[30] As we have seen, the tenor of the whole document was to affirm the unity of humankind and the monogenist account. Whether or not monogenism was specifically asserted, rather than just assumed, was a matter of considerable debate, but most commentators agreed that polygenism was condemned. Just what was to be made of this rejection of polygenism was not so clear, however. Augustin Bea, for example, believed that the encyclical did not address the *scientific* side of the issue at all, whatever it affirmed for theology. He felt that whether there were forms of polygenism that could be made consonant with Church-teaching was a question that had actually been shelved.[31] And then other observers pointed out that the way in which polygenism was outlawed still left room for preadamites; still others felt that even if this were so, it mattered little, for preadamism was merely an antiquated seventeenth-century theory.[32] Time and again, however, the ritual rejection of preadamism in one mode was only prefatory to its reassertion in another. Thus among those individuals partial to evolutionary science, the preadamite theory was invariably invoked to validate the thesis that truly human types had developed from sub-human preadamic stock.

A sophisticated treatment of this possibility from the perspective of Catholic philosophical theology is to be found in the writings of Karl Rahner. In Rahner's mind preadamism, in the sense of fully human individuals existing before Adam, was so entwined with polygenism that he felt compelled to reject it, although he made it clear from a detailed scrutiny of the Church's official pronouncements that, in his judgment, polygenism was

---

[30] An introduction to some of the immediate commentaries at the time in English, French, German, Spanish and Italian is provided in Gustav Weigel, "Gleanings from the Commentaries on Human Generis," *Theological Studies*, 12 (1951): 520-549. Among those dealing specifically with scientific aspects of the encyclical were G. Vandebroek and L. Renwart, "'Humani Generis' et les Sciences naturelles," *Nouvelle Revue Théologique*, 73 (1951): 3-20; Guy Picard, "La Science Expérimentale est-elle Favorable au Polygénisme?" *Sciences Ecclesiastiques*, 4 (1951): 65-89.

[31] Augustin Bea, "Die Enzyklika 'Humani Generis': Ihre Grundgedanken und ihre Bedeutung," *Scholastik*, 26 (1951): 36-56.

[32] See, Weigel, "Gleanings" (cit. n. 30), 544. Still, even those who gave a scientific assessment of the question of the unity of the species still reviewed the early preadamite theory of Peyrère. See J. Carles, "Polygénisme ou Monogénisme. Le Problème de l'Unité de l'Espèce Humaine," *Archives de Philosophie*, 17 (1954): 84-100.

not explicitly ruled out. What was crucial here, both to Rahner and to the spirit of Catholic officialdom, was a doctrine of original sin umbilically tied to "natural, biological generation" from the first ancestor—a conception quite at variance with Calvinist federal theology.[33] On this account any acceptable polygenism would of necessity implicate its advocates in a theology of multiple original falls-from-grace. Thus for Rahner determining a viable stance on the *scientific* question of human origin was inextricably bound up with the theological conception of original sin. At the same time, it also depended on whether the human constitution was to be conceived of dualistically (as body and soul) or as fundamentally indivisible. Following the current drift of the theological tide, both Protestant and Catholic, Rahner expressed his reservations about crudely dualistic models that smacked of Greek infiltration.[34]

With these twin commitments to original sin and to human psychosomatic unity, Rahner insisted that "in regard to the animal kingdom man is a metaphysically new, essentially diverse species, not merely in the biological sense of the phenotype, not only in name, but in that ultimate root of his psychosomatic nature." And yet such "creationist" talk was to him entirely compatible with a belief that "biological development in the animal kingdom reached so advanced a stage of development in a number of exemplars that the transcendent miracle of 'becoming man' could take place in them."[35] What precisely Rahner meant by this assertion is not entirely clear, for he certainly did not envisage any naive plopping of a soul into an animal body. But he had in mind the "hominization" of preadamic creatures nonetheless.

---

[33] Karl Rahner, "Theological Reflexions on Monogenism," in *Theological Investigations* (cit. n. 19), 229-296, on p. 244. It should be noted, however, that Rahner did mention the possibility of a theory of original sin "without a generative connexion"—"not indeed generatione, but per inoboedientiam primi hominis, non imitatione." ibid., 270-271. Teilhard de Chardin's theory of anthropogenesis, according to T.V. Fleming, also viewed Adam as the juridical father of the human race and thus allowed for the possibility of polygenism. In this scenario the leap from animal to human, which produced Adam, required divine intervention. See T.V. Fleming, "Two Unpublished Letters of Teilhard," *The Heythrop Journal* 6 (1965): 36-45.

[34] In general terms, dualists found it easier to work with a preadamic perspective on human origins, since Adam's creation could be conceived of as the "ensoulment" of a prehuman creature. Protestant resistance to this, and predictably to Peyrère's preadamism, is evident in G.C. Berkouwer, *Man: The Image of God*, translated by Dirk W. Jellema (Grand Rapids: Eerdmans, 1962), 280. A move back towards duality is apparent in Robert H. Gundry, *Soma in Biblical Theology with Emphasis on Pauline Anthropology* (Cambridge: Cambridge University Press, 1976), and in John W. Cooper, *Body, Soul, and Life Everlasting: Biblical Anthropology and the Monism-Dualism Debate* (Grand Rapids: Eerdmans, 1989). For a Catholic perspective on these issues see Russell Coleburt, "The Special Creation of the Soul," *Downside Review*, 90 (1972): 235-244.

[35] Rahner, "Theological Reflexions on Monogenism" (cit. n. 19), 294-295.

What is significant *is* that when he came to teasing out just how the "creation of the spiritual soul" was accomplished, he resorted again to those embryological analogies already so serviceable to Catholic evolutionists:

If the mediaeval doctrine is presupposed, and it is coming to the fore again, that the spiritual soul only comes into existence at a later stage in the growth of the embryo, several pre-human stages will lie between the fertilized ovum and the organism animated by a spiritual soul. These do not yet, therefore, stand in immediate and proximate potency to actuation by the spiritual soul. . . . On that basis it is quite possible to say that an ontogeny viewed in that way corresponds to human phylogeny as present-day evolutionary theory sees it. In both cases a not yet human biological organism develops towards a condition in which the coming into existence of a spiritual soul has its sufficient biological substratum.[36]

With this is mind it is not at all surprising that Rahner should approvingly have cited Philipp Dessauer's comments that "shortly before the appearance of man, simian forms very close to man" had emerged. With manual freedom, upright posture, and human teeth, they represented a "group of prehuman, animal forms" whose purpose was "to prepare the way for man."[37]

Even more recently the question of the relationship between polygenism and Catholic theology was the subject of a dissertation by Augustine Kasujja. Reviewing in some detail Peyrère's original preadamite theory and linking it with the polygenism of Voltaire, Calhoun, Nott, and Gliddon, Kasujja moved on to discuss the significance of the discovery of fossil humans for the Catholic understanding of the origin of the human species. To advance his case Kasujja drew on East African archaeological discoveries, in particular *Homo habilis*, to focus the question whether monocentrism or polycentrism provided the best scientific explanation of human origins. What if, he pondered, advocates of polycentrism were to be vindicated? Would the Magisterium and Scripture allow for polygenism? For him the key theological question centered on the transmission of original sin, and he argued that if this were to be conceived of in terms of solidarity rather than genetic descent—just as some Calvinist theologians had suggested—then the door would be opened to a reconsideration of

[36] Karl Rahner, *Hominization. The Evolutionary Origin of Man as a Theological Problem* (New York: Herder and Herder, 1965), 93-94.

[37] Rahner, "Theological Reflexions on Monogenism," (cit. n., 19), 295. See also E.L. Mascall's Bampton Lectures for 1956, *Christian Theology and Natural Science* (London: Longmans, Green & Co., 1956), 254-289 on "Man's Origin and Ancestry." Rahner's polygenetic inclinations are also evident in his "Evolution and Original Sin," in Johannes Metz (ed.), *Consilium. Volume 26. The Evolving World and Theology* (New York: Paulist Press, 1967), 61-73 in which he considers that it would be "better if the magisterium refrained from censuring polygenism," 64.

polygenism as a viable Catholic option.[38] Clearly the Catholic tradition, despite appearances to the contrary, has not as yet given its last word on polygenetic accounts of human evolution.

Similar theological maneuvers are also detectable within certain strands of mainstream Protestant theology, despite surface appearances to the contrary. Take for example Emil Brunner. To him, the whole enterprise of harmonizing the modern world view of science with that of biblical cosmogony is a profoundly misguided project, and a *fortiori* the attempt to integrate biblical anthropology with the science of prehistoric archaeology a myopic undertaking. "To equate the Neanderthal Man," he writes, "with 'Adam in paradise'—an attempt which is being made to-day, even by European theological university professors—merely produces an impossible bastard conception, composed of the most heterogeneous and incongruous views." And yet for all his denials, Brunner remained convinced of a fundamental distinction between *animalitas* and *humanitas*: "*animalitas* here denotes a form of existence which achieves no personal acts at all, but merely acts of self-preservation and the preservation of the species; *humanitas* means, however, that in which—even if at first in a very rudimentary way—something personal, something which transcends *animalitas*, is achieved." Thus, when Brunner asserts that "even if man is descended from the animal world, as *humanus* he is something wholly new, not only in contrast to the ape, but over against the whole of Nature," we see the perpetuation of the essential threads of evolutionary preadamism woven into the fabric of neo-orthodox theology.[39]

Despite these observations of Brunner's, it is clear that the survival of any theologically significant version of the preadamite theory is, by and large, restricted to evangelical Protestants and orthodox Catholics. As I have said, the reason for this is simply because the idea only retains power—both scientifically and theologically—for those for whom the doctrine of a historic Adam has significance. And thus it is, that the very theory that began its life in skepticism now finds itself called upon to preserve theological orthodoxy in the face of latter-day scientific challenges.

---

[38] Augustine Kasujja, *Polygenism and the Theology of Original Sin Today. Eastern African Contribution to the Solution of the Scientific Problem: The Impact of Polygenism in Modern Theology* (Rome: Urbaniana University Press, 1986). I am grateful to Bram Wiedenaar for drawing this dissertation to my attention.

[39] Emil Brunner, *The Christian Doctrine of Creation and Redemption. Dogmatics* volume II, translated by Olive Wyon (London: Lutterworth, 1955), 50, 81.

# Conclusion

As a means of harmonizing the Christian religion with the claims of science, preadamism has proved to be a remarkably versatile strategy. We have seen, for example, how its early deployment as a source of skeptical biblical criticism later gave way to its adoption by advocates of racial ideology to justify a white theodicy from both polygenetic and monogenetic perspectives. Preadamism, moreover, possessed the resources to meet challenges from evolution theory, and this has facilitated its continued survival into the twentieth century.[1] Besides these scientific functions it independently raised important questions within philosophical theology, about the concept of the soul, the *imago dei*, and the fall of Adam.

All of this serves to remind us that harmonizing strategies—like science and religion themselves—truly are cultural resources. The preadamite theory, accordingly, was a resource for certain religious believers to preserve the coherence of their own views of the world. Indeed, as we have seen, unbelievers could resort to the preadamite theory not only as a source of skepticism but as a means of charging believers with heresy! As often as not, when "conflict" did occur in the discourse of science and religion, it was between the naturalistic commitments of scientific skeptics and the harmonizing tactics of believers. Moreover, in different contexts preadamism performed different social functions. Some, like Winchell, could use monogenetic preadamism for racist ends, while Warfield employed the same scheme for quite the opposite. Whereas Poole's polygenetic version was thoroughly humanitarian in thrust, Fleming's was just as thoroughly elitist. To say that preadamism subserved social functions, is not to prejudge the issue as to which social ends it advanced.

In an important sense, preadamism acted to keep the peace between science and religion. More than that, it performed the role of a kind of conceptual bridge between the two spheres and transformed each in the light of the other. To put it another way,

---

[1] Indeed with the current interest in human origins stimulated by the search for so-called "Mitochondrial Eve" and the monogenetic "Noah's Ark Theory" of human evolution, a revival of some version of preadamism among conservative Christians seems rather likely.

preadamism, once espoused, interrogated both scientific theory and theological conviction. Scientifically, preadamism—depending on the precise version adopted—predisposed its advocates to certain explanatory alignments. They frequently found themselves committed to particular theses about linguistic origins, about the role of environmental modification in organic history, and about the fixity of type. By the same token, it also had religious implications for the Christian doctrines of the fall, the image of God, the soul, and biblical hermeneutics. Harmonizing strategies, it seems, are thus rarely single-unit ideas; rather, they are conceptual systems—packages of ideas—that transform the very notions they seek to unite.

Finally, a word or two about the relationship between preadamism as a harmonizing strategy and the secularization of scientific theory seems appropriate. Initially, and ironically, preadamism, which began life as a secularizing trend in scientific endeavor, by challenging the detailed accuracy of biblical chronology and by liberating anthropological investigation from Mosaic strictures, ended up by being baptized as a reconciling tactic. Here the contingency of theological labeling, from heresy to orthodoxy, is all too plainly exposed. The reason for this change in status, moreover, has to do with the secularization of scientific explanation at the macro-level. As theological talk was more and more filtered out of general scientific discourse, the preadamite theory increasingly operated at the micro-level of accommodating theology to particular empirical questions. Thus, while in its earlier day the language of preadamism was good currency within the arena of science, it progressively became restricted either to internal theological dialogue or to works of Christian apologetics. Again the irony is plain: the very ideas that aided the secularization of science in the first instance became rehabilitated within the more restricted confines of the theological world.

Preadamism, clearly, has served different cognitive purposes. And it has served different social groups as well. Perhaps it is because it could fulfill such wide-ranging needs—from the skeptical to the apologetic, from humanitarianism to elitism—that it continues to lurk in the cultural memory of the Christian West, periodically resurfacing to serve some scientific, theological, or social interest.

# INDEX